Student Solutions Manual

Interactive Statistics

Third Edition

Martha Aliaga
University of Michigan

Brenda Gunderson
University of Michigan

PEARSON

Prentice
Hall

Upper Saddle River, NJ 07458

Editor-in-Chief: Sally Yagan
Executive Editor: Petra Recter
Supplements Editor: Joanne Wendelken
Executive Managing Editor: Kathleen Schiaparelli
Assistant Managing Editor: Becca Richter
Production Editor: Allyson Kloss
Supplement Cover Manager: Paul Gourhan
Supplement Cover Designer: Joanne Alexandris
Manufacturing Buyer: Ilene Kahn

© 2006 Pearson Education, Inc.
Pearson Prentice Hall
Pearson Education, Inc.
Upper Saddle River, NJ 07458

Printed in the United States of America

10 9 8 7 6 5 4 3 2 1

ISBN 0-13-149837-1

Pearson Education Ltd., *London*
Pearson Education Australia Pty. Ltd., *Sydney*
Pearson Education Singapore, Pte. Ltd.
Pearson Education North Asia Ltd., *Hong Kong*
Pearson Education Canada, Inc., *Toronto*
Pearson Educación de Mexico, S.A. de C.V.
Pearson Education—Japan, *Tokyo*
Pearson Education Malaysia, Pte. Ltd.

Contents

1.1

In hypothesis testing, the purpose is to determine whether there is sufficient evidence with which to reject the null hypothesis (H_0), which generally reflects the prevailing viewpoint. The alternative hypothesis (H_1) is often what someone is hopeful that the data will support.

1.3

H_0: The 5-year survival rate for all those using the vaccine is equal to 10%.
H_1: The 5-year survival rate for all those using the vaccine is greater than 10%.

1.5

(a) Since a Type I error is rejecting H_0 when H_0 is true, we would be concluding the gun is not loaded when it is loaded. Since a Type II error is failing to reject H_0 when H_1 is true, we would be thinking the gun is loaded when it is not. A Type I error may be more serious as one might accidentally shoot a loaded gun.

(b) Since a Type I error is rejecting H_0 when H_0 is true, we would be concluding the dog does not bite when it does bite. Since a Type II error is failing to reject H_0 when H_1 is true, we would be thinking the dog bites when it does not. A Type I error may be more serious as we might approach a dog that could bite.

(c) Since a Type I error is rejecting H_0 when H_0 is true, we would be concluding the mall is closed when it is open. Since a Type II error is failing to reject H_0 when H_1 is true, we would be thinking the mall is open when it is closed. A Type II error may be more serious as we might waste the time and gas to drive to the mall expecting it open when it is closed.

(d) Since a Type I error is rejecting H_0 when H_0 is true, we would be concluding the watch is waterproof when it is not. Since a Type II error is failing to reject H_0 when H_1 is true, we would be thinking the watch is not waterproof when it is waterproof. A Type I error may be more serious as we might ruin our watch if it gets wet.

1.7

H_0: The average tomato yield for Brand A fertilizer is the same as the average tomato yield for the more expensive Brand B fertilizer.
H_1: The average tomato yield for the more expensive Brand B fertilizer is greater than the average tomato yield for the Brand A fertilizer.
Type I Error: Spend more money on the Brand B fertilizer when it really is not better than the Brand A fertilizer regarding the average tomato yield. Type II Error: Continue to use the Brand A fertilizer when the Brand B fertilizer results in a higher tomato yield on average.

1.9

A Type I error is rejecting the null hypothesis when it is true. So the owner would conclude the patrons are older and the owner would spend the time and money to remodel, when the crowd is actually not older. The owner would have spent money unnecessarily and the remodeling may not appeal to some of the patrons, but in general, it is not a serious error.

1.11

(a) The null hypothesis could not be rejected.
(b) No, a complaint was not registered.
(c) Yes, a Type II error may have been made. The cans are thought to contain the stated sodium content when actually they contain higher amounts of sodium on average.

1.13

If α is decreased then β will increase, so the possible value is 0.30.

1.15
(a) The significance level α is 2/30=0.067 and the level of β is 20/30=0.667.
(b) Decision Rule #2: Reject H_0 if the selected voucher is $\leq$ \$2 or is $\geq$ \$9. The significance level α is 6/30=0.20 and the level of β is 12/30=0.40. Enlarging the rejection region resulted in increasing the level of α from 0.067 to 0.20 while decreasing the level of β from 0.667 to 0.40.

1.17
No, we need a decision rule that states when we reject or fail to reject H_0.

1.19
(a) False: $\alpha + \beta$ does not need to equal 1. The value of α is calculated under H_0 while the value of β is calculated under H_1
(b) False: Type II error is the chance of failing to reject H_0 when H_1 is true.
(c) True.
(d) False: H_0 is rejected if the sample shows evidence against it.
(e) False: The sample size does not influence the alternative hypothesis. The alternative hypothesis can be one-sided no matter what the sample size.

1.21
(a) H_0: The shown box is Box A. H_1: The shown box is Box B.
(b) The direction of extreme is one-sided to the left.
(c) Reject H_0 if the selected token is \$5 or less.
(d) The significance level $\alpha = 2/25 = 0.08$ which is less than 0.10.
(e) The chance of a Type II error is $\beta = 11/25 = 0.44$.
(f) Our decision is to reject H_0.

1.23
(a) False. $\alpha + \beta$ does not need to equal 1. The value of α is calculated under H_0 while the value of β is calculated under H_1.
(b) False. A Type II error occurs if H_1 is true and we fail to reject H_0. Here we don't know if H_1 is true.
(c) True.

1.25
(a) The frequency plots are provided below:

```
        Bag X                                    Bag Y
          X                         X                              X
          X                         X                              X
    X     X     X                   X                              X
    X     X     X                   X                              X
    X     X     X                   X                              X
    X     X     X                   X                              X
    X     X     X                   X     X                  X     X
    X     X     X                   X     X                  X     X
 X  X     X     X     X             X     X     X      X     X
Blue Brown Yellow Green Red        Blue Brown Yellow Green Red
```

(b) No, the response being recorded is the color, which has no particular ordering for the outcomes. If the colors had been listed in the order of Blue, Red, Brown, Green, and Yellow, then the apparent direction of extreme would be one-sided to the left. So it is not appropriate to discuss a direction of extreme in this case.

1.27
The p-value should be small in order to reject the null hypothesis H_0. A small p-value indicates that the observed data or data even more extreme is very unlikely or unusual if the null hypothesis is true. In general, we reject H_0 if p-value is less than or equal to α, the significance level.

1.29

(a) Frequency plots for the two completing hypotheses.

H_0:

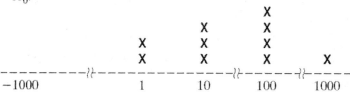

```
                                    X
                        X           X
            X           X           X
            X           X           X
- - - - - - - ?? - - - - - - - - - - - ?? - - - - - ?? - - - -
  -1000          1          10        100       1000
```

H_1:

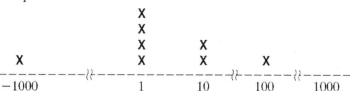

```
                    X
                    X
                    X           X
    X               X           X
- - - - - - - ?? - - - - - - - - - ?? - - - - - - ?? - - - -
  -1000          1          10        100       1000
```

(b) α = chance of a Type I error = chance of rejecting H_0 when H_0 is true
= chance of observing \$1 from the winning bag
= 2/10 = 0.20

(c) β = chance of a Type II error = chance of failing to reject H_0 when H_1 is true
= chance of observing \$10 or \$100 from the losing bag
= 3/8 = 0.375

(d) No, we did not actually observe a voucher.

Box I

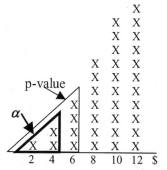

1.31

(a) The direction of extreme is one-sided to the left (to the smaller values).

(b) The chance of a Type I error is $\alpha = 1/6 = 0.1667$ (found under the Die A model for 1 or less). The chance of a Type II error is $\beta = 7/10 = 0.70$ (found under the Die B model for 2 or more).

(c) The p-value is the chance of getting the observed value of 2 or less, assuming the die is Die A, so the p-value = 2/6 = 0.333. Since this p-value is greater than the significance level α of 0.1667, we fail to reject H_0 and conclude that the selected die appears to be Die A.

1.33

(a) H_0: The shown bag is Bag A. H_1: The shown bag is Bag B.

(b) The direction of extreme is two-sided.

(c) (i) The p-value is $4/40 = 0.10$.

(ii) Yes, since the p-value is $\leq \alpha$.

(iii) No, since the p-value is $> \alpha$.

(d) (i) The p-value is 1.

(ii) No, since the p-value is $> \alpha$.

(iii) No, since the p-value is $> \alpha$.

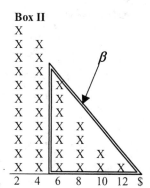

1.35

(a) All new model 100-watt light bulbs produced at Claude's plant.

(b) H_0: The population of all new model 100-watt light bulbs (produced at Claude's plant) has an average lifetime equal to 40 hours. H_1: The population of all new model 100-watt light bulbs (produced at Claude's plant) has an average lifetime greater than 40 hours.

(c) 10%

(d) The p-value can be any value between 0 and 0.10.

(e) Yes, if the p-value is less than or equal to 0.10, then the p-value is also less than or equal to 0.15 so it is significant at the 0.15 level. However, if we only know that the p-value is less than or equal to 0.10, we cannot be sure whether the p-value is also less than or equal to 0.05. Without further information about the value of the p-value we cannot determine if the data would also be significant at the 0.05 level.

1.37

(a) H_0

(b) p-value > 0.10

(c) We failed to reject H_0, so we could have made a Type II error.

(d) One-sided to the right. We want to see if the numbers have increased.

1.39

(a) We rejected H_0, so we could have made a Type I error.

(b) We decide that the average cost is higher than \$350 while it is really not. Maybe you decide that the cost is too high and decide to attend a different college, while in reality you could have attended this college after all.

(c) The p-value is ≤ 0.10.

(d) Yes.

1.41

(a) For study A, a possible p-value is 0.001; for study B, a possible p-value is 0.11; and for study C, a possible p-value is 0.03.

(b) Reject H_0 if the p-value is small, so support for H_0 is shown if the p-value is large, the largest p-value is for Study B.

(c) We rejected H_0, but it was true, so a Type I error was made.

(d) For Study A: one-sided to the right, Study B: two-sided, Study C: one-sided to the left.

1.43

(a) See the chart below for the alternative hypotheses.

(b) See the chart below for the possible p-values.

(c) The results for Study C had the most support for the null hypothesis since the p-value was the largest.

(d) This would be called a Type I error.

	Null Hypothesis	Alternative Hypothesis	*p*-value
Study A	The true proportion of females is equal to 0.60.	The true proportion of females is not equal to 0.60.	0.08
Study B	The average time to relief for all Treatment I users is equal to the average time to relief for all Treatment II users.	The average time to relief for all Treatment I users is less than the average time to relief for all Treatment II users.	0.005
Study C	The true average income of adults who work two jobs is equal to \$70,000.	The true average income of adults who work two jobs is greater than \$70,000.	0.20

1.45

(a) True.

(b) False.

1.47

(c) to be statistically significant at the 5% level means the p-value is less than or equal to 0.05. However, we do not know if the p-value is less than or equal to 0.01 or if it is between 0.01 and 0.05, so the answer is "sometimes yes" (if the p-value is also ≤ 0.01) and "sometimes no" (if the p-value is > 0.01).

1.49

(a) You observed a yellow ball, which is the most extreme result that you could get. With only one observation there is no more extreme than observing a yellow. So the p-value is the chance of observing a yellow ball under the null hypothesis, which is $1/5 = 0.20$. Since the p-value is larger than 10%, the result is not statistically significant.

(b) The data consists of selecting two balls with replacement (and the order is not important). The possible outcomes are shown in the picture below:

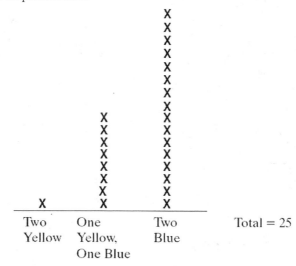

We observed one yellow and one blue ball. Results that are even more extreme would be observing two yellow balls. So the p-value is $(1+8)/25 = 0.36$. Since the p-value is larger than 10%, the result is not statistically significant.

1.51

Statement (a) since the effect is small so we would need a larger sample size to detect it.

1.53

(a) (i) α = chance of observing an average ≥ 45 from bag A = $(4+2+1)/190 = 0.0368$

    (ii) 45

    (iii) $\beta = 0.2789$

    (iv) The chance that we decide that the bag shown is bag A, while it really is bag B is 0.28 (or 28%).

(b) (i) First note that the observed average voucher value is $35, so we have: p-value = chance of observing an average of $35 or more extreme under H_0 = $(17+9+4+2+1)/190 = 0.1737$

    (ii) p-value $\leq \alpha$ so we reject H_0

    (iii) $20+17+9+4+2+1/190 = 0.2789$

    (iv) p-value $> \alpha$ so we fail to reject H_0

    (v) The cut-off value is the average of $35, since we rejected H_0 for $35 and we failed to reject H_0 for $30.

1.55

(a) 20 nCr 2 = 190

(b) 20 nCr 3 = 1140

1.57

(a) The sketches are provided below.

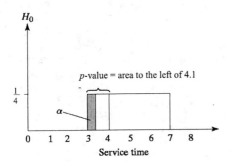

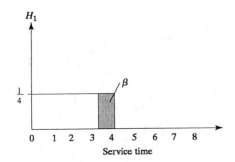

(b) (i) See the shaded area marked as α above.

(ii) $\alpha = (0.4)(1/4) = 0.1$

(iii) See the shaded area marked as β above.

(iv) Power $= 1 - \beta = 1 - (0.6)(1/4) = 1 - (6/10)(1/4) = 1 - 0.15 = 0.85$.

(c) (i) See the shaded area marked as the p-value above.

(ii) p-value $= (1.1)(1/4) = 0.275$.

(d) No, the p-value $> 0.1 = \alpha$.

2.1
35% is a parameter – it is a numerical summary of the population.
28% is a statistic – it is a numerical summary of a sample from the population.

2.3
The proportion of register voters in Ann Arbor who would vote "Yes" on Proposal A is an example of (c) a parameter.

2.5
For a simple random sample of $n = 200$ and the number of items that were defective = 5. All we can say is that (c) the percent of defective items in the sample is 5/200 = 2.5%. $\hat{p} = 5/200 = 0.025$ or 2.5%.

2.7
(a) The value of 34 is a statistic.
(b) Response bias.

2.9
Response bias.

2.11
Nonresponse bias.

2.13
The best answer is (c).

2.15
(a) Yes, each sample of size 20 has the same chance as any other sample of size 20 to be selected (assuming all tags are exactly the same and the box is thoroughly mixed).
(b) It is drawn without replacement.

2.17
(a) H_0: The population proportion of dissatisfied customers equals 0.10.
     H_1: The population proportion of dissatisfied customers is less than 0.10.
(b) With the calculator the selected customers are: 34318, 15553, 8461, and 614. With the random table the selected customers are: 15409, 23336, 29490, and 30414.
(c) Type II error.
(d) 0.21.
(e) No, the p-value is > 0.05.
(f) (ii) Statistic.

2.19
Stratified random sampling. You are dividing up your population by class rank then selecting 100 students from within each stratum at random.

2.21
(a) 45 8ᵗʰ-grade students; 25 10ᵗʰ-grade students; 48 12ᵗʰ-grade students. For a total in the sample of 118 students.
(b) With the calculator the selected students are: 193, 127, 430, 100, and 427. With the random number table the selected students are: 241, 304, 22, 364, and 151.

2.23

We should use a weighted average: $\left(\dfrac{40}{100}\right)(70) + \left(\dfrac{60}{100}\right)(63) = 65.8$ inches

2.25

(a) Stratified random sampling.
(b) With the calculator the first five selected homes are: 386, 81, 379, 211, and 156. With the random number table the first five selected homes are: 94, 299, 396, 378, and 363.
(c) $(0.60)(2100) + (0.40)(2600) = 2300$ square feet.

2.27

(a) Stratified random sampling.
(b) With the calculator the first six homes selected from the 1000 homes in County I are: 918, 193, 902, 502, 370, and 5. With the random number table the first six homes selected from the 1000 homes in County I are: 963, 19, 197, 705, 463, and 79 where the homes are labeled 0 to 999.
(c) $(0.50)(175) + (0.30)(200) + (0.20)(195) = 186.5$ thousands of dollars or $186,500.

2.29

(a) With the calculator and the first 100 addresses labeled 1 through 100, the sample is: Addresses = #79, #179, #279, #379, #479. Using the random number table with the first 100 addresses labeled 01, 02, ... , 98, 99, 00, the sample is: Addresses = #30, #130, #230, #330, #430.
(b) $1/100 = 0.01$ or 1% There are 100 possible systematic samples of size 5, each equally likely.

2.31

Yes, the chance is 1/5 or 0.20 or 20%. There are 5 possible systematic samples, each equally likely.

2.33

(a) $1/8 = 0.125$.
(b) Note that there are 350 members so 350/8 → 43.75 or 43 groups of 8 and one last group of 6. For the 300th member to be selected, the starting point must have been a 4, that is, the 4th member in each group of 8 is selected. This will result in a sample of 44 members.

2.35

False, the chance depends on the number of clusters.

2.37

(a) Cluster sampling. The faculty are grouped into departments which serve as clusters. Six of the clusters are selected at random. All of the units in the cluster are in the sample.
(b) Label the list of the 60 departments 1, 2, 3, ..., 60. The numbers generated and thus the departments selected using the TI calculator with seed = 79 are: 54, 37, 49, 5, 15, and 43. If you are using the random number table, you might label the list of the 60 departments 01, 02, 03, ... , 60. The numbers generated and thus the departments selected starting at row 60, column 1 are: 23, 22, 47, 40, 25, and 37.
(c) Yes we can determine the chance that any specific professor will be selected. The chance of selecting a professor is the same chance that his/her department or cluster will be selected. The reason being if a cluster is selected, every element in the cluster is selected. Therefore the probability that a cluster is selected is 6/60 = 0.10, the chance that any specific professor will be selected.

2.39

(a) The type of sampling performed in each dorm is cluster sampling, with the rooms forming the clusters and 3 clusters were selected at random from each of the four dorms.
(b) If each cluster selected has one student, we would have 3 students from each of the four dorms for a minimum sample size of 12 students.
(c) If each cluster selected has three students, we would have 9 students from each of the four dorms for a maximum sample size of 36 students.

(d) We do not know, it will depend on how many clusters selected are rooms with women. The number of women could be as low as 0 and as high as 36.

(e) No, there will be anywhere from 3 to 9 freshmen students sampled from the freshmen dorm, as well as from 3 to 9 sophomores, from 3 and 9 juniors, and from 3 to 9 seniors.

2.41

(a) With the calculator or the random number table, the selected region is 3 = Southwest.

(b) Stratified random sampling.

(c) (i) 1-in-10 systematic sampling.

(ii) 0.10.

(iii) With the calculator, the first five selected cans are 7, 17, 27, 37, and 47. With the random number table we might label the first can 1, the second can 2, ..., and the 10th can 0. Then the first five selected cans are 7, 17, 27, 37, and 47.

(iv) Note that 125/10 = 12.5 → 12 or 13 cans. However, there will not be a 7th can to select in last group. Thus the total number of cans in the sample will be 12.

(d) (i) No.

(ii) Two possible values are 0.12 and 0.15.

(iii) Yes, a Type II error.

2.43

(a) A cluster sampling of blocks.

(b) Yes, the chance is 5/50 = 0.10.

(c) No, we do not know how many households are in each block. Additionally, if the number of households in each block was not the same across all blocks, then we would also need to know which of the 5 blocks were selected.

(d) With the calculator, the five selected blocks are 8, 44, 33, 6, and 38. With the random number table we might label the blocks 01 to 50, then the five selected blocks are 12, 18, 11, 43, and 05.

(e) Response bias because the interviewers are college students. People may not feel comfortable telling these students they want to forbid loud music at parties in the college dorms.

2.45

The proportion 153/200 is a parameter since the instructor polled her entire class and that was the group that she was interested in learning about.

2.47

Nonresponse bias.

2.49

This survey may be subject to nonresponse bias. Only those alumni who respond and report their income will be included. Alumni who perhaps are currently unemployed or in a low paying position may elect not to respond. Therefore, the reported average income based on such a survey may be biased upwards -- the average may be larger than the actual average for all alumni.

2.51

(a) With the calculator the selected ID numbers are: 179, 2274, and 3327. With the random number table the selected ID numbers are: 2398, 2258, and 3540.

(b) A statistic.

2.53

(a) Based on her decision rule, Jane rejects H_0.

(b) Since Jane rejected the null hypothesis, the data are statistically significant.

(c) No, based on a sample of two $1 Jane can not be certain of which purse she has since both purses contain at least two $1.

(d) Type I error: reject H_0 when H_0 is true.

(e) Simple random samples of size 2 from null purse:
   $\$1_1$ and $\$1_2$, $\$1_1$ and $\$5_1$, $\$1_1$ and $\$5_2$
   $\$1_2$ and $\$5_1$, $\$1_2$ and $\$5_2$, $\$5_1$ and $\$5_2$

(f) Simple random samples of size 2 from alternative purse:
   $\$1_1$ and $\$1_2$, $\$1_1$ and $\$1_3$, $\$1_1$ and $\$1_4$
   $\$1_2$ and $\$1_3$, $\$1_2$ and $\$1_4$, $\$1_3$ and $\$1_4$

(g) The p-value is the chance of getting two $1 bills or more extreme (in the direction of H_0, but in this case, there is no "more extreme") if H_0 is true. The p-value is the chance of getting two $1 bills if H_0 is true, i.e., 1/6.

2.55
Stratified random sampling. You are dividing up your population by gender then selecting your sample within the strata at random.

2.57
(a) All former university graduate students.
(b) Stratified random sampling.
(c) False.

2.59
(a) Stratified random sampling
(b) The chance is 0, only 1 of the two fiction books (A, B) will be selected.
(c) Two books that could be selected are Book A and Book C.
(d) Another pair of books that could be selected is Book A and Book D.
(e) The chance that the total number of pages exceeds 800 is the same as the chance that the two selected books are Book B and Book D. There are 4 possible pairs of books that could be selected of which the (B, D) pair is 1, so the chance is $1/4 = 0.25$.

2.61
(a) With the calculator the first student selected is 1. With the random number table the first student selected is 1.
(b) The sample size is 6.

2.63
(a) (i) $0.02(500)+0.03(1200)+0.05(18000) = 10+36+900 = 946$.
   (ii) Stratified Random Sampling.
   (iii) With the calculator the selected labels are 432, 232, 304, 412, and 372. With the random number table, we might assign the first High category driver the labels 001 and 501. We would assign the second High category driver 002 and 502. This assignment pattern would continue until the 500th High category driver who would be assigned the labels 000 and 500. Reading off labels from row 60, column 1, we have: 789, 191, 947, 423, and 632. This would correspond to selecting the 289th, 191st, 447th, 423rd, and 132nd High category drivers in the list of 500 High category drivers.
(b) (i) With the calculator or the random number table, the first selected label is 15. Thus the selected labels will be 15, 35, 55, 75, 95, and so on.
   (ii) Since the 500 High category drivers divide evenly into groups of 20 ($500/20 = 25$), there will be a total of 25 High category drivers in the systematic 1-in-20 sample.

2.65
These results are based on a study of 125 aerobic classes in five health clubs, not all aerobic classes in all health clubs. Thus, the 60% figure is a statistic and the sample size is $n = 125$.

2.67
(a) Cluster sampling.
(b) 1/5.
(c) Response bias.

2.69

(a) A 1-in-40 systematic sample.

(b) 1/40 is the chance that any specific address is chosen since one of the first 40 addresses is picked at random. The other addresses are directly linked to that first random pick.

2.71

(a) Multistage, with the first stage being a cluster sample of 3 lab sections and the second stage being a simple random sample of 25% of the students in the 3 selected labs.

(b) 0.0625.

(c) The 78% is a statistic since it was based on the sample of students surveyed.

2.73

(a) Since 15 patients were selected with a 1-in-9 systematic sample, there were at least 15 groups of 9 patients each or 135 patients in all.

(b) (ii) Statistic.

2.75

(a) Answers will vary. See the web site.

(b) Summaries will vary.

3.1
(a) Explanatory variable: Amount of coffee consumed.
    Response variable: Exam performance.
(b) Explanatory variable: Hours of counseling per week.
    Response variable: Grade point average.
(c) Explanatory variable: Number of roommates.
    Response variable: Physical health at the end of the semester.
(d) Explanatory variable: Type of driver (levels are good or bad).
    Response variable: Reaction time on a driving test.

3.3
(a) Experiment.
(b) The response variable is improvement in blood flow. The explanatory variable is the amount of flavonoid (High versus Placebo).
(c) p-value ≤ 0.05
(d) The 178.8 is a statistic and a sample mean.

3.5
(a) Observational study.
(b) Blood cholesterol level.
(c) Frequency of egg consumption.
(d) Diet, age, amount of exercise, and history of high cholesterol are a few possible confounding variables.

3.7
(a) Observational study.
(b) Homelessness.
(c) Some are: separation status (whether or not they were separated from their parents), poverty Status (whether or not they were raised in poverty), family problem status (whether or not they have family problems), abuse status (whether or not they were sexually or physically abused).
(d) All homeless people in Los Angeles.
(e) The homeless people in Los Angeles that were surveyed.
(f) The 2.5% figure is a statistic since it is based on a sample of homeless people in Los Angeles.

3.9
(a) This is a retrospective observational study.
(b) The response variable is car crash status. The explanatory variable is gender.
(c) We should take into account that men typically drive more than women. In this case, "miles driven" is a confounding variable that can skew your results. Another confounding variable could be weather conditions at the time of the crash.

3.11
(a) H_1.
(b) p-value was less or equal to 0.05.
(c) (iii) Observational study, Prospective.

3.13
(a) Observational study (retrospective).
(b) Response variable is kidney stones status (whether or not the subject develops kidney stones); Explanatory variable is amount of calcium in diet.
(c) A diet high in calcium lowers the risk of developing kidney stones in men and women.

3.15
(a) Observational study (retrospective).
(b) Response variable is Alzheimer's disease status, explanatory variable is linguistic ability.
(c) The chance of low linguistic ability given a person has Alzheimer's.

3.17
(a) It is an experiment because the various treatment combinations were actively imposed on the experimental units.
(b) The experimental units are the metal clutches.
(c) The response variable is the lifetime of the clutch.
(d) The factors or explanatory variables are   type of oven and temperature. We have 4 types of oven and 3 levels of temperature.
(e) There are 12 treatments.
(f) We would need 24 clutches, 2 at each of the 12 treatments.
(g) A design layout table for this experiment is given as:

Factor 1: Type of Oven

		Type 1	Type 2	Type 3	Type 4
Factor 2: **Temperature**	Temp 1	2 clutches	2 clutches	2 clutches	2 clutches
	Temp 2	2 clutches	2 clutches	2 clutches	2 clutches
	Temp 3	2 clutches	2 clutches	2 clutches	2 clutches

3.19
(a) Durability.
(b) Dye color and type of cloth.
(c) 20
(d) 20 x 6 = 120.

3.21
(a) Batches of feed stock.
(b) Yield of the process.
(c) Temperature (2 levels) and Stirring Rate (3 levels).
(d) 6 treatments.
(e) 24 units.
(f) Design Layout Table:

Factor 1: Temperature

		50	70
Factor 2: **Stirring** **Rate**	60	4 batches	4 batches
	100	4 batches	4 batches
	140	4 batches	4 batches

3.23
(a) Weight.
(b) Fiber content and carbohydrate content.
(c) Fiber levels (3) Low, Medium and High. Carbohydrate levels (2) Low and Medium.
(d) 3 x 2 x 20 = 120.

3.25

Answer is (d).

3.27

The rats are labeled 1 through 8. Use a calculator with seed= 1209 or a random number table with row=24, column=6 to select the 4 rats to receive the treatment. The other 4 rats will not receive the treatment. At the end of the week the effectiveness of the vaccine against the virus will be measured.

3.29

(a) Experiment.
(b) Cause of death.
(c) Estrogen therapy with two levels: receive the treatment, do not receive the treatment.
(d) · Placebo.
(e) H_0: Women who receive Estrogen therapy do not have a lower risk of dying from a heart disease than women who do not receive the Estrogen therapy

H_1: Women who receive Estrogen therapy have a lower risk of dying from a heart disease than women who do not receive the Estrogen therapy.
(f) The p-value was less than or equal to 0.05.
(g) Yes, Type I error.

3.31

(a) With the calculator the selected rats were: 14, 13, 5, 16, 9, 3, 4, 2, 7, 10. With the random number table the selected rats were: 4, 13, 8, 5, 12, 15, 2, 6, 9, 20.
(b) Have another researcher make and record the measurements.

3.33

Blocking can help to reduce the bias due to possible confounding variables that you know of and thus build into the design of the experiment. Randomization can then be used to help reduce the bias due to other possible confounding variables that you do not know of or have not measured (sometimes these are referred to as lurking variables).

3.35

Answers will vary.

3.37

Answers will vary based on the available case studies and the one selected.

3.39

Observational study. No active treatment was imposed.

3.41

(a) Temperature (3 levels) and Baking Time (2 levels).
(b) Taste.
(c) 6 treatments.
(d) 36 batches.

3.43

Lack of blinding of the subjects and of the experimenter(s) or evaluator(s).

3.45

(a) Experiment.
(b) The response variable is skin rash status. The explanatory variable is skin rash treatment.
(c) Using the calculator, the selected labels are: 10, 14, 23, 12, 11, 18, 3, 25, 26, 22, 27, 24, 13, 2, and 4. Using the random number table, the selected labels are: 09, 26, 06, 17, 13, 25, 18, 10, 16, 19, 24, 28, 14, 30, and 02.
(d) The 20% value is a statistic.
(e) Two possible values are 0.08 and 0.10. Any two values larger than 0.05 (but less than 1) would work.

3.47
(a) It is not clear from the article.
(b) It may not be appropriate to extend these results to the general male population, nor to the female population.
(c) The results are more meaningful when they can be adjusted for such factors. Confounding variables.

3.49
(a) Observational study.
(b) Attitude towards mathematics.
(c) Gender.
(d) Cluster sampling.
(e) Selection bias.

3.51
(a) iii. confounding variable.
(b) i.   False
    ii. False
    iii. True
    iv. False

3.53
The placebo effect is a phenomenon in which receiving medical attention, even administration of an inert drug, improves the condition of the subjects.

3.55
(a) Experiment.
(b) Proportion of juice added to the drink.
(c) Rating of the juice on a scale of 1 to 100.
(d) iii. Confounding variable.
(e) ii. Single-blinded.
(f) With the calculator, all five volunteers received 5% juice.
    With the random number table, the first four volunteers received 10% juice and the last one received 5% juice (using 0-4 to represent 5%, using 5-9 to represent 10%).
(g) False. The probability of rejecting the null hypothesis when it is true is 1%.

3.57
(a) Experiment.
(b) Using the calculator: With seed = 23 the selected men were: BJ(3), AB(13), JP(4), ZB(15), TD(18), CF(16), MG(6), TG(17), MK(2). With seed= 41 the selected women were: NL(11), BG(2), MM(5), MA(3), AG(9), KS(6), JB(4), NI(16), SL(10).
    With the random number table: Men(row 12, column1): RM(9), TG(18), MK(2), VN(8), LG(12), SK(5), WB(11), BJ(3), RM(9). Women(row 22, column 1): NI(16), KB(17), MA(3), SM(7), BG(2), KS(6), CI(15), SL(10), MM(5).
(c) Confounding variable.

3.59
(a) Using the calculator with seed= 45, the first five subjects selected are: 3351, 3140, 860, 703, and 5488.
    With the table of random numbers (row 10, column 5) the first five subjects selected are:
    5368, 5753, 3425, 3988, and 5306.
(b) The placebo effect is a phenomenon in which receiving medical attention, even administration of an inert drug, improves the condition of the subjects.
(c) 3.5% of 4058 = 142.
(d) Statistic.
(e) H_0
(f) p-value > 0.05.
(g) Type II error.

3.61

(a) Observational study.

(b) We don't know whether the patients in both groups were in the same physical condition.

(c) Assign to the 100 patients numbers 1-100. With the calculator or with the random number table select at random the first 50 patients. Those patients will receive treatment 1, the other patients will receive treatment 2.

(d) No. The treatment is an operation.

3.63

(a) The treatment of type of tanning bed was actively imposed (assigned) by the experimenter.

(b) It is not entirely clear by the description provided in the article. If randomization was used, then the completely randomized design seems most likely, randomly selecting some of the 14 to receive the UV filter and the remaining to not.

(c) The response variable is a mood assessment score. The explanatory variable is the UV status with levels of UV rays or No UV rays (i.e. filter).

(d) With the calculator the seven subjects assigned to the UV filter bed would be: 5, 7, 11, 6, 9, 2, and 12. With the random number table and labeling the 14 subjects as 01 to 14, the seven subjects assigned to the UV filter bed would be: 09, 06, 13, 10, 14, 02, and 04.

(e) Yes, the control was the tanning bed with the UV filter. It only states that the subjects were *directed* to one of the two seemingly identical tanning beds. We do not know if randomization was used. The subjects were blinded since the two beds were 'seemingly identical'. If each of the two tanning beds were used by at least two subjects, then replication was incorporated.

(f) There are a number of possible caveats. This study was based on volunteers. Was the researcher who made the mood assessments blinded with respect to the treatment each volunteer received? Did the UV filter have any effect on the perceived quality of the tan received? If so, perhaps those with the filter were not as satisfied with the results and thus less content. The sample size is somewhat small.

3.65

(a) Observational study.

(b) Explanatory variable: Amount of wine.
Response variable: IQ score.

(c) Amount of wine was likely self-reported and may result in a response bias.

(d) Who paid for the study? Who were the subjects on the study? How many subjects were selected?

3.67

(a) Experiment.

(b) Explanatory variable: Behavioral therapy status.
Response variable: Insomnia status.

(c) The number of people in each treatment group is unknown. We only know that they were 75 in total.

(d) How were the results analyzed?
How were the 75 subjects selected? Was the placebo treatment truly a placebo?

3.69

(a) Experiment.

(b) The response variable is the difference between the estimated (perceived) relative ability and the actual ability. The explanatory variables are skill level and difficulty of task.

(c) There is no information about the number of subjects or how the subjects were selected. There is very little detail about the actual design of the experiment. It states that the subjects were grouped according to their skill level, but it does not say how this was done. There is no information on how the ability (perceived or actual) was measured. Finally, the last three paragraphs bring up an important concept called interaction that will be revisited in Chapter 12 on analysis of variance.

3.71

(a) There are 10 subjects with Dry skin, 6 subjects with Normal skin, and 4 subjects with Oily skin. Form blocks of size 2 within each skin type. The first two subjects with Dry skin type would form the first pair, the next two selected at random from the remaining four would form the second pair, and so on. This would be followed within each skin type group.

(b) The 10 subjects with Dry skin have the following ordered ages: 31, 32, 43, 44, 51, 53, 60, 61, 64, and 64. So blocks of size 2 would be formed using this age order. The first block would consist of the 31 and 32 year old subjects with dry skin; namely, volunteers 13 and 14. The second block would consist of the 43 and 44 year old subjects with dry skin; namely, volunteers 1 and 8. The other three dry skin blocks would be: volunteers 3 and 7, volunteers 16 and 20, and volunteers 5 and 12. Likewise, we form blocks of 2 by ordering the 6 normal skin type subjects by their ages: 20, 21, 25, 26, 56, and 58. Resulting in the following blocks of size 2: volunteers 6 and 18, volunteers 2 and 15, and volunteers 9 and 17. The ordered ages for the four subjects with oily skin type are: 27, 29, 47, and 48. Resulting in the following blocks of size 2: volunteers 4 and 19, volunteers 10 and 11.

(c) No, they should go to each block of 2 subjects and randomly select one to receive the new facial lotion treatment. The randomization should be conducted within each block, not across the blocks. Then the responses for the two subjects within each block would be compared to assess the difference in the two treatments.

4.1
(a) qualitative
(b) quantitative-continuous
(c) qualitative
(d) quantitative-discrete
(e) qualitative
(f) quantitative-continuous

4.3
(b) qualitative.

4.5
The total number of students who failed the midterm may be a valid measure of the difficulty of the class. In addition to that information, one needs information regarding the total students in each class. There may be only 20 students in the English class but a total of 200 students in the Chemistry class. A more useful measure may be the proportion or percent of students who failed the midterm. Other questions that one may ask: What type of students takes each class? Is the class a requirement?

4.7

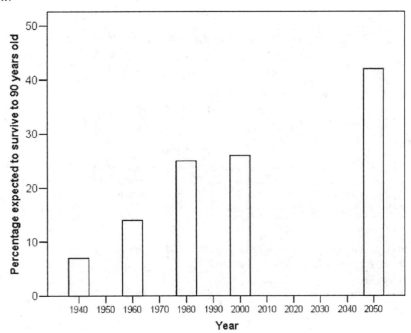

4.9

(a) The sales are represented by the areas of the bars, not just the height of the bars.

(b)

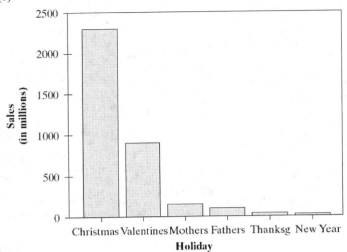

4.11

(a) $0.20(620) = 124$

(b) No, background is a categorical variable. It is not appropriate to discuss shape in a bar graph since the order of the categories is arbitrary.

4.13

(a) The conditional distribution of years of experience given gender is:

		Years of Experience		
		1-6	*7-12*	*13+*
Gender	*Male*	25.7%	37.9%	36.4%
	Female	42.3%	35.1%	22.6%

(b) It appears that male coaches generally have more years of coaching experience than females. However, there have been more women (in terms of percentages) entering the field of coaching. Since Title IX went into effect, and the interest in women's sports is rising, many new coaching jobs have been available for women's sports. So a high percent of relatively *new* female coaches is reasonable.

4.15

(a) Men: $16,120/(16,120+1,728+1,708+3,751+1,462) = 16,120/24,769 => 65.08\%$
    Women: $2,406/(2,406+670+522+810+1633) = 2,406/6,041 => 39.83\%$

(b) The conditional distribution is:

		Method of Suicide				
		Firearms	*Drugs*	*Hanging*	*Gases*	*Other*
Gender	*Men*	65.1%	5.9%	15.1%	6.9%	7.0%
	Women	39.8%	27.0%	13.4%	8.6%	11.1%

(c) The bar chart is shown below:

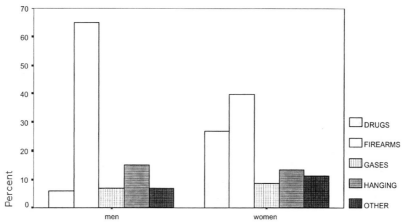

(d) Men who commit suicide overwhelmingly choose firearms while women are more evenly split between firearms and drugs.

4.17

(a) 36/50 = 0.72 or 72%

(b) 24/50 = 0.48 or 48%

(c) Yes. The p-value for the test of no association is 0.01431, which is less than the 0.05 level of significance.

4.19

(a) The table below is one way to present the information.

Number of high-achieving students assigned		Number of low-achieving students assigned	
3 years in a row to ineffective teacher	3 years in a row to effective teacher	3 years in a row to ineffective teacher	3 years in a row to effective teacher
30	80	77	40

(b)

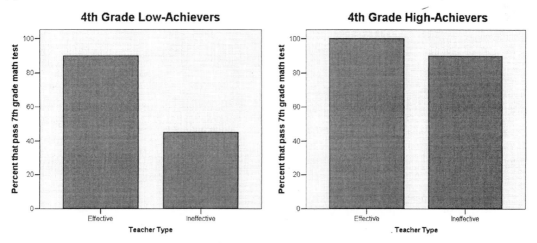

Number of Low-Achievers Assigned to Teacher Type

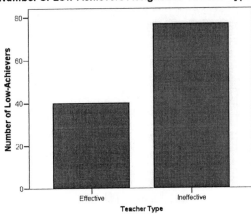

Number of High-Achievers Assigned to Teacher Type

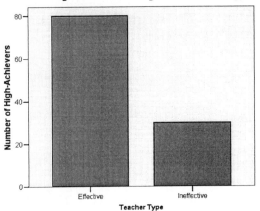

4.21
Answers will vary.

4.23
Stem-and-leaf plot: Note: 83 | 4 represents an octane rating of 83.4.

```
83 | 4
84 |
85 |
86 | 7
87 | 5 7
88 | 0 4 5 8
89 | 1
90 | 5
```

4.25
(a) Back-to-back stem-and-leaf plot. Note: | 4 | 5 represents a score of 45 points.

```
Student B          Student A
             4 | 5 8
             5 | 2
             6 | 9
  8 6 5 4 3  7 | 6 9
  7 4 2 1 0  8 | 0 6
             9 | 2 4
```

(b) Student A's scores range from 45 to 94. Student A did receive the highest score of the two students. However, Student A also received the lowest score of the two students. The scores for Student B ranged from 73 to 87. Student B performed very consistently in the 70's and 80's.

4.27
(a) 781 hours (a Longlife light bulb).
(b) If you wanted the brand that gives the longest lasting light bulb.
(c) If you wanted the brand that lasts longest on average (that is using the mean).

4.29

(a) Note: 80|8 represents an On-Time departure rate of 80.8%.

% On-Time			% On-Time
Arrival	6 4	75	Departures
	0	76	
	8 1 0 0	77	
		78	
	6	79	
	1	80	8
	5 2 1 1	81	
	5	82	
	1	83	1 3 8 9
	9 1	84	3
	8	85	7 7
	4 0	86	1 5 6
	9	87	0 6
		88	1 6 8 9
		89	0 8 9
	3	90	
		91	3
		92	6

(b) The arrivals range from 75.4% to 90.3%. The arrival values are quite dispersed with a gap on the high side between Detroit (87.9%) and Raleigh (90.3%). The departures range from 80.8% to 92.6%. There is less variation among the departures as compared to the arrivals. The departures distribution is somewhat bimodal with a gap on the low side between Chicago (80.8%) and Denver (83.1%).

(c) Airports perform better overall with respect to on-time departures. The center value for on-time departures (86.8%) is higher than the center value for on-time arrivals (81.15%). The range for the on-time departures is smaller than for the on-time arrivals. With the exception of Chicago, all of the % on-time departures are higher than the center value for the % on-time arrivals.

4.31

(a) $3 + 6 + 8 + 10 + 5 = 32$.

(b) $9/32 = 0.2813$

4.33

(a) The distribution is skewed to the right.

(b) $5/31 \Rightarrow$ about 16%

(c) Since 5 minutes corresponds to 300 seconds, we have 2 people.

(d) We cannot state the maximum length of time a customer had to spend in line. All we know is that a customer spent between 350 and 400 seconds in line but we don't know the actual time.

4.35
Back-to-Back Histograms

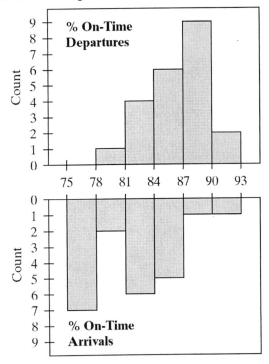

Note: If you are using the TI calculator, you might set the window as follows: min= 75, max = 95, Xscl = 3, Ymin = -1, Ymax = 10, Yscl = 1, and Xres = 1.

4.37
(a) For the end of 1925, the shape is somewhat uniform but with an extreme modal class of 100+. For the end of 1960, the shape is positively skewed (or skewed to the right), with the most frequent price class of 20-25. For the end of 1994, the shape is skewed to the right with the most frequent price class of 10-15.
(b) The general shift has been from many stocks priced 100+ to more stocks priced between 5-40.
(c) Yes. A lot of stock splits would lead to more stocks at lower prices.

4.39
Time Series I matches Description (1) since we would expect the proportion of babies born that are girls to fluctuate around 0.50. Time Series II matches Description (2) since cumulating the number of batteries that fail would increase over time.

4.41

(a) Time plot.

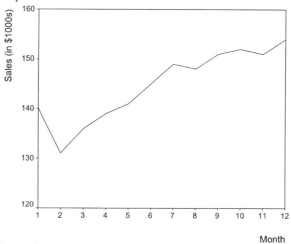

(b) an increasing trend.

4.43

(a) 25% National, 29% Michigan

(b) 18% Michigan, 17% National

(c) That information is not given. It is not true that since 25% of the fatal accidents nationally involved female drivers, then 75% of the fatal accidents involved male drivers because a male and female driver may both die in a crash.

(d) In 1983, both the Michigan and the National percentages were around 20%. For Michigan that was a decrease from the previous year, however, nationwide, this was a slight increase from 1982. Both for the Michigan and the National numbers, the 1984 numbers were up from 1983. In Michigan there was a slight dip in the numbers in 1983, nationwide there was not.

(e) More women are working; consequently, more women are driving.

4.45

(a) Plot C

(b) Negative linear association.

4.47

Let x be the independent variable and y be the dependent variable

(i) (a) x = Year; y = Population of the U.S.

   (b)

   (c) The population of the U.S. is increasing

(ii) (a) x = Time; y = Temperature of a drink

(b)

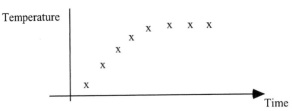

(c) The temperature increases at first and after a while levels off to approximately room temperature.

4.49

Let x = hours of exercise per week, and y = level of cholesterol. As you increase the number of hours of exercise, (hopefully) your level of cholesterol will decrease. Another example is: Let y = the temperature of a cup of hot tea outside the house in the winter; and x = the length of time it is outside the house.

4.51

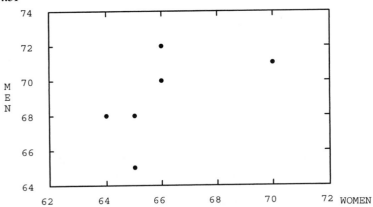

The scatterplot shows a positive association between male height and female height. The linearity might be reasonable but it is not a very strong linear association. Note that for five out of the six couples, the man was taller than the woman. It may be appropriate to superimpose a 45 degree line on the graph if the question of similar refers to having the same height.

4.53

(a)

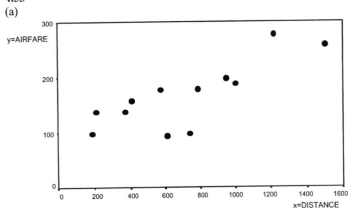

(b) Yes, positive.
(c) Fairly strong.

26 *Chapter 4*

4.55
(a) Median income decreased to nearly $36,000.
(b) Median income increased from $35,000 to nearly $40,000.
(c) Median income decreased back to approximately $37,000.
(d) Predications would be difficult without further information to perhaps help explain the changes described in parts (a), (b), and (c).

4.57
(a) You might ask: Is this an average SAT score? What data were used to get this average? Is the spending an average amount spent? Across all schools? Just some schools? Both large and small schools?
(b) The actual SAT score range is: 400 - 1600. With the updated graph the increase in spending and decrease in SAT would be less dramatic.

4.59
(a) quantitative continuous
(b) quantitative continuous
(c) quantitative continuous
(d) quantitative discrete
(e) qualitative
(f) quantitative discrete
(g) quantitative continuous

4.61
(a) We have 4179+3766+251 = 8196 students.
(b) 88 of the 251 students, or 88/251 => 35.06%.
(c) The pie chart would look the same with 35.06% for Hispanic, 35.06% for Black, 17.1% for White, and 6.8% for Asian. You might also include an "Other" category with the remaining 5.98%.

4.63
Answer is: (d) The conditional distribution of Blood Type given Rh Factor.

4.65
(a) Observational study.
(b) Color blindness status (with responses being yes or no).
(c) Gender (with levels being male or female).
(d) Statistic (2% is a sample percentage).
(e) Stratified with female = stratum I and male = stratum II.
(f) (i) Yes; (ii) Can't tell; (iii) No
(g) The answer is (ii).

4.67
(a) Time is continuous, but since it was recorded to the nearest 5 minute interval it is discrete.
(b) The distribution is skewed to the right with an outlier at 105 minutes. The waiting times range from 5 minutes to 105 minutes. The median is 32.5 minutes.
(c) Since the distribution of waiting times is skewed, the mean waiting time is not an appropriate measure of the typical waiting time. The median is resistant to outliers and skewed distributions. Therefore, the median (32.5 minutes) is an appropriate typical waiting time.
(d) It does not say how that day was selected or whether the 30 patients were all of the patients on that day or a sample.

4.69
.(a) Amount of Radiation Emitted (mR/hr).
(b) No. We only know that the range falls between 0.1 and 0.9. We don't know the exact upper and lower limits of the range.
(c) There are 10 out of 20 stores or 50%.
(d) Since 0.5/0.08 = 6.25, the maximum we could use is 6 small television sets.

4.71

(a) Time plot.

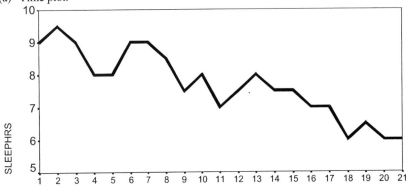

DAY

(b) (v) decreasing trend.

4.73

(a) Time plot.

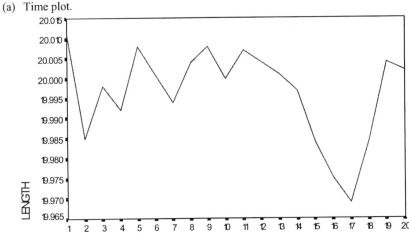

Sequence number

(b) Observations 1 through 13 seem to be varying around the target value of 20.000 cm. Observations #16 and #17 are extremely low.

(c) What happened after observation #13? Did the operator fall asleep? Did the settings get bumped?

4.75

(a) Scatterplot.

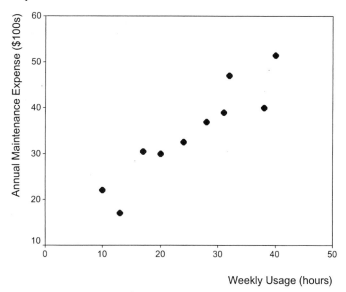

Weekly Usage (hours)

(b) Yes, the relationship does appear to be linear.
(c) The relationship appears to be positive.
(d) There are no unusual values or outliers.

4.77

(a) A person born in 1946 would be 19 years old in 1965 and a person born in 1964 would be 1 year old (or younger) in 1965.
(b) Some of the 1,867,000 people being tracked over time have died.
(c) No, there heights of the bars are not given, nor any axis with values to be able to approximate the heights.
(d) The distribution is skewed to the right for both 1965 and 1995, with the skewness being stronger in 1965. In projected 2025 the distribution is nearly uniform.

4.79

(a) The enforcement officers in Utah were assaulted more often between 10 PM and 2 AM and less often between 6 AM and 8 AM.
(b) The second figure shows the continuity of time and better depicts the most common times for assaults to occur.
(c) How was "assault" defined? Were these Utah data based on all cities, counties, rural areas?

Summarizing Data Graphically 29

5.1
(c) mean

5.3
The overall average for the 40 students would be: $[25(82) + 15(74)]/40 = 79$.

5.5
The median may have been preferred if the distribution of resale house prices was skewed to the right. The median is a more resistant measure of center, and would not be so greatly affected by the few very expensive homes.

5.7
(a) Known: $\dfrac{X_1 + X_2 + X_3 + X_4 + X_5 + X_6 + X_7 + X_8 + X_9 + X_{10}}{10} = 35$,

thus, $X_1 + X_2 + X_3 + X_4 + X_5 + X_6 + X_7 + X_8 + X_9 + X_{10} = 350$

now enter $X_{11} = 32$, the new mean is :

$\dfrac{X_1 + X_2 + X_3 + X_4 + X_5 + X_6 + X_7 + X_8 + X_9 + X_{10} + X_{11}}{11} = \dfrac{350 + 32}{11} = 34.727$

(b) We cannot find the new median because we do not know where the new age will fall in the order of ages.

5.9
(a) Yes, I agree. The mode is the value occurring most frequently.
(b) For the values: 1, 1, 1, 2, 20; the median = 1 = minimum and the mean = 5. So 4 out of 5 or 80% of the values are below the mean

5.11
(a) (i) $s = 9.016$
   (ii) $s = 0$
   (iii) $s = 2.669$
(b) The standard deviation is so large due to one very large outlier, namely the value of 27. Extreme deviations from the mean greatly inflate the standard deviation.
(c) (i) range $= 27 - 1 = 26$
   (ii) range $= 7 - 7 = 0$
   (iii) range $= 11 - 3 = 8$
   The range is misleading for data set (i). If the outlier of 27 were removed, the range would be 8.

5.13
(a) From the boxplot, the highest start-up cost is about $670,000.
(b) From the histogram, the distribution appears to be skewed to the right.
(c) The mean, since the boxplot shows that the median is less than $100,000.
(d) It is the largest value that is not considered a potential outlier, the largest observation that falls within the upper inner fence.

5.15
(a) True.
(b) True.
(c) False.
(d) True.
(e) False.

5.17

Statement (b).

5.19

(a) Min = 28, Q1 = 32, Median = 40, Q3 = 48, Max = 50.5

(b) 15

(c) 31

(d) IQR = 48 – 32 = 16 years

(e) 1.5(16) = 24; lower fence = 32 – 24 = 8; Since 20 is > 8, the observation would not be an outlier.

5.21

(a) All three are equal, since they all have the same range of 70 – 30 = 40.

(b) Class C.

(c) Class A: Min = 30, Q1 = 50, Median = 50, Q3 = 50, Max = 70.
  Class B: Min = 30, Q1 = 40, Median = 50, Q3 = 60, Max = 70.
  Class C: Min = 30, Q1 = 30, Median = 50, Q3 = 70, Max = 70

5.23

The mean in Celsius would be found as: (5/9)[28-32] = -2.22, the standard deviation in Celsius would be found as: $|5/9|(10) = 5.56$.

5.25

(a) On average the temperatures were about 3.43 °F from their mean of 953.2 °F.

(b) Min = 949, Q1 = 950, Median = 953.5, Q3 = 955, Max = 959

(c) It could increase as much as you want. It will never change the median, since it will still be the middle value.

(d) We can see that there is an upward trend; the temperature seems to increase slowly.

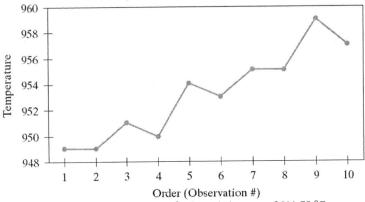

(e) On average the temps were about 1.91 °C from their mean of 511.78 °C.

(f) Min = 509.44, Q1 = 510, Median = 511.94, Q3 = 512.78, Max = 515

5.27

(a) Quantitative Standard Score = -0.41
  Analytical Standard Score = 0.70
  Verbal Standard Score = 1.91

(b) Verbal Comprehension

5.29

We have that [(70)(52)+(30)(lab score 2)]/100 = 64. So lab score 2 = [(100)(64) - (70)(52)]/30 = 92.

5.31

(b) The mean would be higher than the median.

5.33

Yes. For the values 1, 3, 20, the mean = 8 and the standard deviation = 10.44.

5.35

(a) These data appear to be skewed to the left with an outlier at zero.

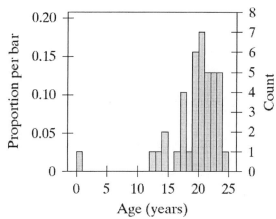

(b) The five-number summary would be preferable since the extreme outlier of zero would influence the mean. Min = 0, Q1 = 17.5, Median = 20, Q3 = 22, Max = 24.

(c) With the outlier removed, the mean and standard deviation may now be appropriate measures to summarize these data. Mean = 19.564, Standard deviation = 2.927.

5.37

Boxplot B. Since the range goes from 2 to 84, the median is below 40, and the shape should reflect a skewed right data set.

5.39

(a) The best performance by the old model resulted in a 60 bottles stuffed in a minute.

(b) Approximately 25% of the time the old model stuffed at least 52 bottles per minute.

(c) On average, the number of bottles stuffed per minute for the new model was 64.05, give or take about 3.15.

(d) Min = 58, Q1 = 62, Median = 64, Q3 = 66.5, Max = 70

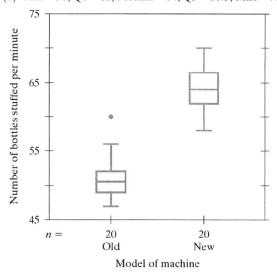

5.41

(a) False: Q1 and the median of 1, 2, 2, 2, 2, 2, 3, 4, 5 are both 2, whereas Q3 is 3.5.

(b) False: If there are 90 boxes that weigh 4 lbs, and 10 boxes that weigh 14 lbs, the total average is [(90)(4)+(10)(14)]/100 = 5 lbs, but there are 90 boxes that weigh less than 5 lbs and only 10 that weigh more than 5 lbs.

(c) False: Both of the following data sets have a mean of 3 and a standard deviation of 1. However, their histograms are not the same. Data set 1: 1, 3, 3, 3, 3, 3, 3, 3, 5; Data set 2: 2, 3, 4.

5.43

(e) none of the above -- From a symmetric boxplot you cannot conclude whether or not its distribution is symmetric so (a) is false. Although the boxplots are exactly the same, the corresponding distributions may not be the same so (b) is false. Since boxplots give the five-number summary and not the average or mean, (c) is false.

5.45

(a) Yes.

(b) Yes.

(c) Happier if the standard deviation is 4, since then your score is 3 standard deviations above the average, a very high score.

5.47

(a) Five-number summary: min=16.55, Q1=26.95, median=45.585, Q3=51.90, max=69.85

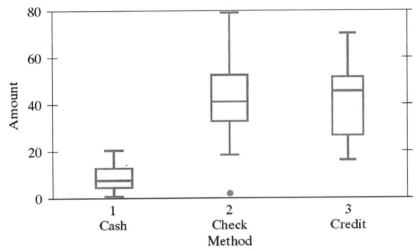

(b) Mean = 41.65, Standard Deviation, s = 14.06

(c) Proportions: cash: 38/100, check: 40/100, credit card: 22/100. About half of the credit card customers came from the customers who paid with cash, and half came from the customers who paid with checks.

(d) From the boxplot, the outlier is at about $1.

(e) Yes, because the cash customers who switch will spend more on average, while check customers who switch will still spend about the same on average.

5.49

The 20 represents 2 standard deviations.

5.51

(a) Time plot.

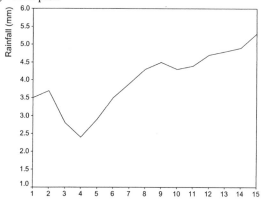

Day

(b) (ii) increasing trend

(c) Min = 2.4 mm, Q1 = 3.5 mm, Median = 4.3 mm, Q3 = 4.7 mm, Max = 5.3 mm.

(d) IQR = 4.7 – 3.5 = 1.2 mm.

(e) (iii)　　the new median = the old median.

(f) (ii) will be larger than the old mean.

5.53

(a) (ii) a prospective observational study

(b) Min = 14, Q1 = 24, Median = 27.5, Q3 = 32.5, Max = 44.

(c) (i) False

　　(ii) Can't tell

　　(iii) Can't tell

　　(iv) True

5.55

(a) Station 2 with a range of 20 – 7 = 13.

(b) Can't tell.

(c) (i) $\bar{x} = 9.67$, s = 3.14

　　(ii) Min = 3, Q1 = 9, Median = 10.5, Q3 = 11, Max = 15

　　(iii) IQR = 11 – 9 = 2; 1.5(2) = 3; Lower fence = 9 – 3 = 6; Upper fence = 11 + 3 = 14. So any observation below 6 or above 14 is a possible outlier, namely, 3, 5, and 15.

　　(iv) Modified boxplot is provided.

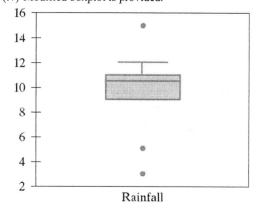

Rainfall

(v) $\bar{x} = 10.67$, s = 3.14.

6.1
They are all the same.

6.3
(a) More than 6.
(b) The proportion of observations to the right of 6 is much larger than the proportion of observations to the left of 6. Since the median divides the total area under the curve in two equal parts, the median must be larger than 6.
(c) The height of the density when $x = 9$ is 1/9. So the area of the two triangles that cover the range from 0 to 9 can be computed as $\frac{1}{2}(6)(1/9) + \frac{1}{2}(3)(1/9) = 0.5$. Thus the area to the right of 9 (that is between 9 and 12) must be $1 - 0.5 = 0.5$.

6.5
(b) The mean of the distribution becomes zero and the standard deviation becomes 1.

6.7
(a) $(680 - 500)/100 = 1.8$.
(b) $(27 - 18)/6 = 1.5$.
(c) Eleanor.

6.9
(a) Prop($Z < 3.49$) = normalcdf (-E99, 3.49,0,1) = 0.9998.
(b) Prop($Z < 4$) = normalcdf(-E99,4,0,1) = 1.
(c) Prop($Z > -0.84$) = normalcdf(-0.84, E99,0,1) = 0.7995.
(d) Prop($Z < 0.84$) = normalcdf(-E99, 0.84,0,1) = 0.7795.
(e) Prop($Z > 2.31$) = normalcdf(2.31,E99,0,1) = 0.0104.
(f) Prop($-3.17 < Z < -1.84$) = normalcdf(-3.17,-1.84,0,1) = 0.0321.
(g) Prop($0.17 < Z < 2.12$) = normalcdf(0.17, 2.12,0,1) = 0.4155.

6.11
(a) Prop($X < 70$) = Prop($Z < (70-63)/5$) = normalcdf(-E99, 70, 63, 5) = 0.9192 or 91.92%.
(b) Prop($X < 70$) = Prop($Z < (70-67)/4$) = normalcdf(-E99, 70, 67, 4) = 0.7734 or 77.34%.

6.13
(a) The mean will be the midpoint of 5 to 9, that is, $\mu = 7$.
(b) By the empirical rule about 68% of the values.

6.15
(a)

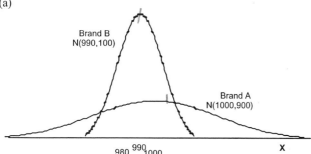

(b) Brand A: 0.7486; Brand B: 0.8413. So a higher proportion of Brand B fuses last longer than 980 days as compared to Brand A.

6.17
Find x, such that Prop($Z < (x-30)/5$) = 0.05. invNorm(0.05,30,5) = 21.78 The maximum duration is 21.78 minutes.

6.19

(a) 0.6915.

(b) 2.8 years.

6.21

(a)

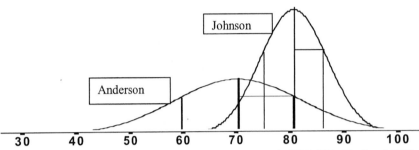

(b) Both give the same. Johnson : Prop(Z > (90-80)/5) = normalcdf(90, E99, 80, 5) = 0.0227.
   Anderson : Prop(Z > (90 -70)/10) = normalcdf(90, E99, 70, 10) = 0.0227.

(c) Johnson : Prop(Z < (50 - 80)/5) = normalcdf(-E99, 50, 80, 5) = 0.00000000099.
   Anderson : Prop(Z < (50 - 80)/5) = normalcdf(-E99, 50, 70, 10) = 0.02275.
   Anderson gives a higher proportion of E's.

(d) Find x such that Prop(Z < (x - 69)/9)= 0.90; invNorm(.90, 69, 9) = 80.53.

6.23

(a) 0.0047.

(b) 56.7 lesions.

6.25

(a) H_0

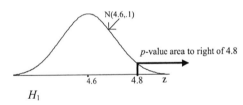

H_1

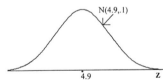

(b) (i) Area to the right of 4.8 under the null hypothesis model corresponds to the p-value.

   (ii) p-value = Prop(Z > (4.8-4.6)/0.1) = normalcdf(4.8, E99, 4.6, 0.1) = 0.02275.

(c) We cannot reject H_0 because the p-value is > 0.01.

6.27

(a) α = Prop(Z < (8 – 15)/3) = normalcdf(-E99, 8,15,3) = 0.0098.

(b) β =Prop(Z > (8 – 10)/3) = normalcdf(8, E99, 10, 3) = 0.7475.

(c) p-value = Prop(Z < (8.5 – 15)/3) = normalcdf(-E99, 8.5, 15, 3) = 0.01513.

6.29

(a) 1.28 standard deviations above the mean for A/B cutoff and
   1.28 standard deviations below the mean for B/C cutoff.

(b) Mean = (50 + 250) /2 = 150 and (250-150)/ σ = 1.28, so $\sigma \approx$ 78.

6.31

(a) invNorm(0.25, 3000, 500) = 2662.76 and invNorm(0.75, 3000, 500) = 3337.24

(b) normalcdf(-E99, 1560.92, 3000, 500) = 0.0020.

(c) We know that 0.002 of N is equal to 200, so N must be equal to 200/0.002 = 100,000.

6.33

(a) $\bar{x}$ = 255.4 grams, s = 2.769 grams.

(b)

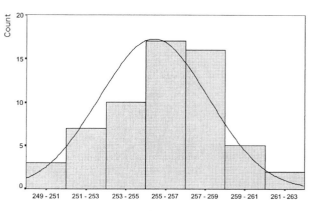

WEIGHT

The distribution appears to be approximately symmetric, bell-shaped, unimodal based on the histogram. The QQ plot below also supports the normal model being appropriate for weight.

Normal Q-Q Plot of Package Weight (grams)

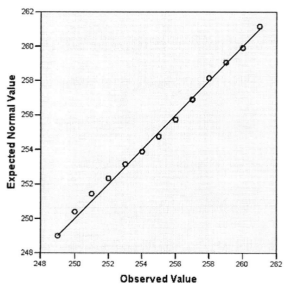

(c) The percentages are 71.7%, 95%, and 100%.

(d) These data appear to resemble a normal distribution approximately.

(e) The product is being filled within the required range approximately 87% of the time.

6.35
(a) The proportion is approximately zero. This could be considered the p-value since it is the chance of the data or more extreme data being observed if the null hypothesis is true.
(b) Proportion $(X < 9) =$ practically zero. We reject H_0 since the p-value is nearly zero. Yes, the result is statistically significant since the p-value is less than any practical choice for α.

6.37
(a) 0.67.
(b) 0.33.

6.39
(a)

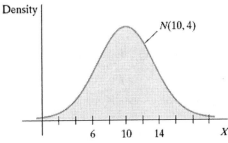 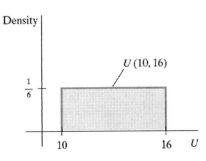

(b) (i) False.
 (ii) True.

6.41
(a)

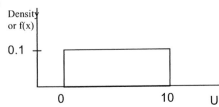

(b) 3/10 or 30%.
(c) i. one-sided to the right.
 ii p-value $= 1/10 = 0.10$
 iii Reject H_0 since p-value $< \alpha = 0.15$.

6.43

(a)

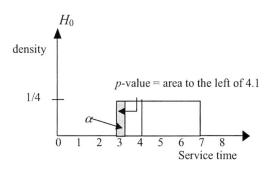

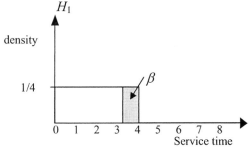

(b) (ii) $\alpha = (3.2 - 3)/4 = 0.05$
    (iv) $\beta = (4 - 3.2)/4 = 0.20$
(c) (ii) p-value $= (4.1 - 3)/4 = 0.275$
(d) No, since the p-value $= 0.275 > \alpha = 0.05$.

6.45

(a) 25%.
(b) More than 1 pound.
(c) More than 1 pound.

6.47

(a) The two missing probabilities are both 0.10.
(b) $0.1 + 0.2 = 0.3$ or 30%.

6.49

0.20.

6.51

(a) $\text{Prop}(X = 0) = \text{Prop}(X = 4) = 0.05$.
(b) $\text{Prop}(X \leq 3) = 1 - 0.05 = 0.95$.
(c) Symmetric, Discrete.

Using Models to Make Decisions 41

6.53

(a)

Y	brown	red	yellow	green	orange	blue
Proportion.	0.30	0.20	0.20	0.10	0.10	0.10

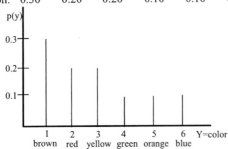

(b) No, because the coding is arbitrary -- if you change the coding, you could change the shape. The variable is qualitative (categorical).

(c) Approximately $1/5 = 0.20$ or 20% should be red.

(d) Take some data (some randomly selected bags). Observe the actual number of each color in the bags. See if the proportions are "close" to those stated.

6.55

Proportion $(X < 36)$ = normalcdf(-E99,36,60,8) = 0.0013.

6.57

(a) Proportion $(64.5 \leq X \leq 72)$ = P((64.5-63.6)/2.5 < Z < (72-63.6)/2.5) = normalcdf(64.5,72,63.6,2.5) = 0.3590.

(b) 1000 (0.359) = 359 women.

(c) 67.7 inches.

6.59

(a) Proportion$(X > 31.5)$ = Prop$(Z > (31.5-30)/3)$ = normalcdf(31.5, E99, 30,3) = 0.3085 or 30.85%.

(b) 34.93 pounds.

6.61

(a) 557

(b) Proportion$(X < 507)$ = 0.6736.

6.63

The mean will be the midpoint of 9.75, and the values of 9 and 10.5 represent approximately 2 standard deviations from the mean, so the standard deviation is approximately $0.75/2 = 0.375$.

6.65

(a) Proportion$(X > 50)$ = Prop$(Z > (50-45.5)/1.5)$ = normalcdf(50,E99,45.5,1.5) = 0.0013.

(b) 0.0015.

(c) 48.58.

6.67

(a) (i) α = normalcdf(13.2, E99, 10, 2) = Prop$(X > 13.2)$ = Prop$(Z > (13.2-10)/2)$ = 0.055

(ii) β = normalcdf(-E99, 13.2, 12, 2) = Prop$(X < 13.2)$ = Prop$(Z < (13.2-12)/2)$ = 0.7257.

(iii) The power of the test = $1 - 0.7257 = 0.2743$.

(b) p-value = normalcdf(11.2, E99, 10, 2) = Prop$(X > 11.2)$ = Prop$(Z > (11.2-10)/2)$ = 0.274.

6.69

(a) p-value = Proportion $(X \geq 62.5)$ = Prop$(Z \geq (62.5-50)/5)$ = normalcdf(62.5,E99,50,5) = 0.0062.

(b) We reject H_0 at the 10% significance level, since the p-value is < 0.10.

6.71

Proportion($X > 20$) = Prop($Z > (20-19)/1.7$) = normalcdf(20,E99,19,1.7) = 0.2782.

6.73

(a) Find x such that Prop($X < x$) = 0.95;   invNorm(0.95, 15, 3) = 19.93 minutes.
(b) Mean of $Y = 1.2\ (15) - 4 = 14$ minutes.
    Standard deviation of $Y = 1.2\ (3) = 3.6$ minutes.
(c) Find x such that Prop($X < x$) = 0.95;   invNorm(0.95, 14, 3.6) = 19.92 minutes.

6.75

(a)

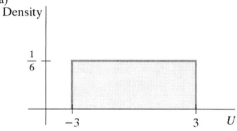

(b) Q1=-1.5 and Q3=1.5, so the IQR=1.5 - (-1.5) = 3. The IQR is not equal to 50%; rather it provides the range of values that cover the middle 50% of the values.

(c)

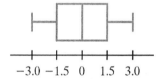

6.77

(a) Proportion($X > 3$) = Prop($Z > (3-1)/2$) = normalcdf(3, E99, 1,2) = 0.1587 or 15.87%.
(b) Proportion($T = 1$) = 0.02, Proportion($T = 5$) = 0.08; Proportion($T > 3$) = 0.28.
(c) Proportion($L > 3$) = 1/3.

6.79

(a) Proportion($X > 3$) = Prop($Z > (3-2)/2$) = normalcdf(3,E99,2,2) = 0.3085.
(b) 0.5.
(d) 17/25.

6.81

The answer is (c), can't tell. We do not know if the distribution is normal.

6.83

(a) 1/6.
(b) 49%.
(c) p-value = 0.5 (1/6) = 1/12 = 0.083.
(d) Since the p-value is < 0.10 , we reject H_0.

6.85

(a)

H_0: X is $U(200,800)$

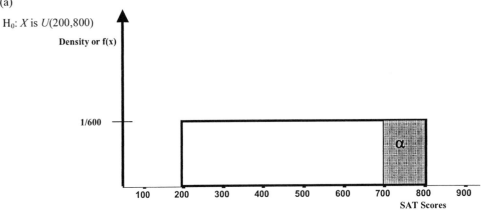

H_1: X is given by:

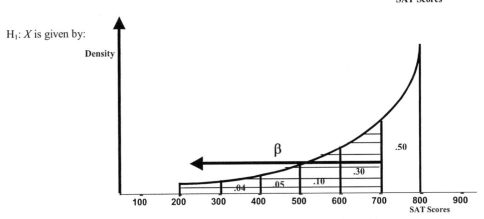

(b) Proportion($X < 300$ under H_1) = $1 - (0.50 + 0.30 + 0.10 + 0.05 + 0.04) = 0.01$.

(c) (i) one-sided to the right.

(d) $\alpha = (800 - 700)/600 = 1/6$ or 16.7%.

(e) β = area to the left of 700 under the H_1 model = 0.50. So the power of the test is $1 - 0.50 = 0.50$.

(f) p-value = $(800 - 400)/600 = 2/3 = 0.67$.

(g) No, because the p-value > α.

44 Chapter 6

7.1
(a) Personal probability.
(b) Relative-frequency approach.

7.3

The long-run relative frequency approach may not be appropriate because not all events are repeatable.

7.5

Answers will vary.

7.7
(a) Mrs. Smith is right.
(b) Answers will vary.

7.9
(a) With the calculator assign 1 as the winning number and 2, 3, 4, and 5 as the losing numbers.
   With the random number table we can assign 01 – 20 as the winning numbers and the other remaining two digit numbers to be the losing numbers.
(b) With the calculator we get the numbers:
   4, 4, 2, 4, 3, 1, 4, 4, 1, 1,   1, 2, 3, 5, 2, 5, 5, 2, 1,5 ,
   5, 2, 1, 1, 2, 2, 5, 4, 4, 4,   5, 4, 5, 3, 2, 1, 4, 4, 1, 3,
   2, 1, 1, 4, 1, 5, 1, 3, 1, 5.
   The first five numbers obtained with the random number table are: 91, 57, 64, 25, 95; and all represent losses.
(c) With the calculator we have: 14/50 = 0.28.

7.11
(a) S = { EEE, EEG, EGE, GEE, EGG, GEG, GG}.
(b) A = {GEG, EGE}.

7.13
(a) $S = \{F, P\}$.
(b) S = {FFFF, FFFP, FFPF, FFPP, FPFF, FPFP, FPPF, FPPP, PFFF, PFFP, PFPF, PFPP,PPFF, PPFP, PPPF, PPPP}.
(c) $S = \{0, 1, 2, 3, 4, 5, 6, 7, 8, 9, 10, 11, 12, 13, 14, 15, 16, 17, 18, 19, 20\}$.

7.15
(a) No, since you can have a king of hearts.
(b) Yes, since you cannot have a card that is both a heart and a spade.
(c) No, since you can have a king of spades.

7.17
(a) Error: 0.19 + 0.38 + 0.29 + 0.15 = 1.01 > 1.
(b) Error: 0.40 + 0.52 = 0.92 < 1.
(c) Error: -0.25 < 0.
(d) Error: Probability is 0 since the events "heart" and "black card" are disjoint.

7.19

(a) 0.70.

(b)

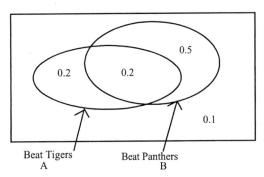

Beat Tigers
A

Beat Panthers
B

(c) 0.90.

(d) 0.20/0.40 = 0.50.

7.21

(a) P(shoes) = 0.52; P(dress) = 0.67; so P(shoes and dress) = P(shoes) P(dress) = (0.52)(0.67) = 0.3484.

(b) P(F)= 0.59; P(at least one F) = 1 − P(no female) = 1 − P(male) P(male)
$$= 1 - (0.41)(0.41) = 1 - 0.1681 = 0.8319.$$

(c) P(French) = 0.78; P(exactly two will have a French citizenship)
$$= 3P(French) P(French) P(no French) = 3(0.78)(0.78)(0.22) = 0.40.$$

(d) $(0.25)^2 (0.75) = 0.0469.$

(e) binomcdf(5 , 0.43 , 3) = 0.88786.

(f) geometcdf(0.38, 4) = 0.85223664.

7.23

(a) P(D) = 110/1160.

(b) P(D | O) = 20/470.

(c) P(O and D) = 20/1160; P(O) = 470/1160; P(D) = 110/1160.
The two events O and D are NOT independent because P(O and D) ≠ P(O) P(D).

7.25

(a) P(Hypertension) = P(H) = 87/180.

(b) P(Hypertension | Heavy smoker) = 30/49.

(c) No, since P(Hypertension and Heavy smoker) ≠ P(Hypertension) P(Heavy smoker)
That is, 30/180 ≠ (87/180) (49/180).

7.27

(a)

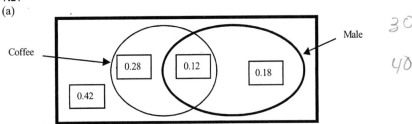

Coffee

Male

30

40

(b) 0.18.

7.29

$1 - (0.18)^2 = 0.9676.$

46 Chapter 7

7.31

$(0.6)(0.6)(0.6) = 0.216$.

7.33

The minimum is 3.

7.35

(a) False.

(b) True.

7.37

(a) No. Here is a counterexample. Suppose the sample space is S={1, 2, 3,4, 5}, and let A= {1} and B= {2} be mutually exclusive. Their complements A^c = {2, 3, 4, 5 and B^c = {1, 3, 4, 5} are not mutually exclusive.

(b) Yes. If A and B are independent then P(A and B) = P(A) P(B). It follows that A^c and B^c are independent since P(A^c and B^c) = P (A^c) P(B^c).

7.39

(a) 1/3.

(b) 0.7143.

7.41

(a) P(I successful) = 0.7, P(II successful) = 0.3
P(introduce new| I successful) = 0.8
P(introduce new| II successful) = 0.4
P(new) = P(new and I) + P(new and II) = (0.7)(0.8) + (0.3)(0.4) = 0.56+ 0.12 = 0.68.

(b) P(I successful | introduce new) = P(I and new)/ P(new) = 0.8235.

7.43

(a) P(I) = 0.7, P(II) = 0.2, P(need repair | I) = 0.03, P(need repair | II) = 0.04,
P(need repair | III) = 0.05,.
P(repair) = (0.7)(0.03) + (0.2)(0.04) + (0.1)(0.05) = 0.034.

(b) Since they are independent (0.034)(0.034) = 0.001156

(c) P(III | repair) = P(III and repair)/ P(repair) = (0.1)(0.05)/0.034 = 0.147.

7.45

(a) P(1) = 0.7, P(D|1) = 0.02, P(D|2) = 0.05
P(D) = (0.7)(0.02) + (0.3)(0.05) = 0.029.

(b) P(1|D) = P(1 and D) / P(D) = (0.7)(0.02)/0.029 = 0.482758.

7.47

(a) 0.2.

(b) 0.1

(c) 0.4.

(d) 0.1/0.4 = 0.25.

7.49

(a) S = {T, WT, WWT}.

(b) X = 1, 2, or 3.

(c) S = {T, WT, WWT, WWWT, …}   so X = 1, 2, 3, 4, …

7.51

(a) P(X = 4) = 0.05.

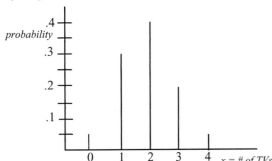

(b) 0.95.

(c) E(X) = 1.9.

7.53

Let X = the amount won (in dollars). Then the distribution for X is shown below.

X = x	2	4	8	....	2^n	...
P(X = x)	1/2	1/4	1/8	...	$1/2^n$	...

So the expected value of X would be computed as: $1 + 1 + 1 + \ldots\ldots = \infty$.

No, it would not be reasonable to pay $1,000,000 to play this game once. However, it would be reasonable to pay it if you could play for as long as you like.

7.55

(a) The expected value for Y is one.

(b) The variance of Y is zero.

7.57

(a) P(X = 4) = 1/6.

(b) P(at least six cars) = 0.833.

(c) E(X) = 6.833.

(d) E(Y) = 2(6.833) – 1 = 12.666.

7.59

(a)

How many subsets contain ...	Answer	The subsets are:	Combination:
... exactly 0 values?	1	empty set	$\binom{5}{0} = 1$
... exactly 1 value?	5	{1}, {2}, {3}, {4}, {5}	$\binom{5}{1} = 5$
... exactly 2 values?	10	{1,2}, {1,3}, {1,4}, {1,5} {2,3}, {2,4}, {2,5}, {3,4} {3,5}, {4,5}	$\binom{5}{2} = 10$
... exactly 3 values?	10	{1,2,3}, {1,2,4}, {1,2,5}, {1,3,4}, {1,3,5}, {1,4,5}, {2,3,4}, {2,3,5}, {2,4,5}, {3,4,5}	$\binom{5}{3} = 10$
... exactly 4 values?	5	{1,2,3,4}, {1,2,3,5}, {1,2,4,5}, {1,3,4,5}, {2,3,4,5}	$\binom{5}{4} = 5$
... exactly 5 values?	1	{1,2,3,4,5}	$\binom{5}{5} = 1$

(b) 32 = 1 + 5 + 10 + 10 + 5 + 1.

7.61

(a) Yes. Each day you win or you don't win, the probability of winning remains the same all the time, and the outcomes are independent of each other.

(b) No. The responses are not independent.

7.63

(a) H_0: 1% of the tax returns were audited, $p = 0.01$

H_1: More than 1% of the tax returns were audited, $p > 0.01$

(b) We have $X \sim \text{Bin}\,(18, 0.01)$, p-value $= P(X \geq 1) = 1 - P(X = 0) = 1 - 0.8345 = 0.1655$

(c) The p-value is the probability of observing exactly one or more tax audited returns among 18, if indeed 1% of tax returns are audited. Or, if we were to take many, many samples of size 18 from this population, we can expect to see 16.55% of them to have 1 or more audited tax returns, if indeed 1% of tax returns are audited.

(d) You cannot reject the null hypothesis.

7.65

(a)

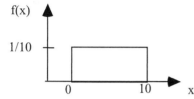

(b) $E(X) = 5$.

(c) 3/10.

(d) (i) One-sided to the right.

(ii) p-value $= 1/10$.

(iii) We reject H_0 since the p-value $< \alpha$.

7.67

(a) X has a Geometric distribution with $P(\text{Thunderstorm}) = p = 0.3$.

(b) $E(X) = 1/p = 1/0.3 = 3.333$.

(c) $P(\text{August 6}) = (0.7)^{36}(0.3) = 7.96\ \text{E-7} = 0.000000796$.

7.69

(a) $P(1050 < X < 1250)$ is the probability that a randomly selected applicant has a SAT score between 1050 and 1250. $P(1050 < X < 1250) = 0.7745$.

(b) (i) 97.5%.

(ii) About 1327.

7.71

(a) $(0.20)(0.70) = 0.14$ or 14%.

(b) $(0.14)(0.15) = 0.021$ or 2.1%.

(c) $0.14 + 0.12 = 0.26$ or 26%.

7.73

(a) $0.12 + 0.10 - 0.17 = 0.05$.

(b) $0.07/0.90 = 0.078$.

7.75

(a) $P(\text{you win}) = 0.10$.

(b) $P(\text{you win}) = 0.41$.

7.77

(a) 0.60.

(b) 0.35.

(c) 0.18.

(d 0.30.

(e) 0.60.

(f) No. $0.18 \neq (0.60)(0.35)$.

7.79

(a) 0.30.

(b) 5/11.

(c) No. The events can occur at the same time.

(d) No. P(Nonsmoker and 0 visit) = 0.25 $\neq$ P(Nonsmoker) P(0 visit) = (11/20)(0.3).

7.81

(a) No, the events are not independent because

$$P(\text{F attend and S attend}) = \frac{18}{80} = 0.225 \neq P(\text{F attend})P(\text{S attend}) = \left(\frac{25}{80}\right)\left(\frac{40}{80}\right) = 0.15625.$$

(b) No, the events are not mutually exclusive because there are 18 families in the intersection.

7.83

The answer is (a).

7.85

(a) Not independent. P(A and B) = 0.1 is not equal to P(A)P(B) = (0.4)(0.3) = 0.12.

(b) Not mutually exclusive, since P(A and B) is not equal to 0.

(c) P(accepted to neither) = 1 – P(accepted to either A or B) = 1 – [P(A) + P(B) – P(A and B)]
 = 1 – [0.4 + 0.3 – 0.1] = 0.40.

7.87

P(positive) = 0.85 and P(negative) = 0.15

P(positive) + P(negative, positive) + P(negative, negative, positive)
 + P(negative, negative, negative, positive) $= 0.85 + (0.15)(0.85) + (0.15)^2 (0.85) + (0.15)^3 (0.85)$
 = 0.99949.

7.89

Let M1 = first machine, M2 = second machine, M3 = third machine, and D = defective item.
$P(\text{M1}) = 0.50$, $P(\text{M2}) = 0.30$, $P(\text{M3}) = 0.20$, $P(D \mid \text{M1}) = 0.04$, $P(D \mid \text{M2}) = 0.06$, and $P(D \mid \text{M3}) = 0.02$.
So $P(D) = (0.50)(0.04) + (0.30)(0.06) + (0.20)(0.02) = 0.042$.

7.91

(a) False.

(b) True.

(c) False.

(d) False.

7.93

(a) $P(X = 4) = 0.05$.

(b) 0.85.

(c) 0.25.

7.95

(a) 0.80.

(b) $E(X) = 2.4$.

7.97

(a)

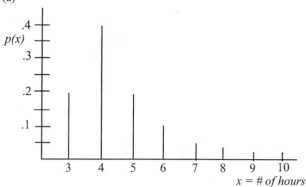

x = # of hours

(b) Mode = 4.
(c) E(X) = 4.58.
(d) Less than.
(e) 0.70.
(f) 0.75.
(g) Response bias.

7.99

(a) 0.60.
(b) E(X) = 3.19.

7.101

Let X represent the diameter, then X has a N(0.8 , 0.02) distribution.
P(not defective) = P(good) = P(G) = P(0.8 − 0.025 < X < 0.8 + 0.025)
                 = normalcdf(0.775 , 0.825, 0.8, 0.02) = 0.7887.
So we have a geometric distribution with parameter p = 0.7887.
P(Def, Def, Def, Def, Def, Good) = $(0.2113)^5(0.7887)$ = 0.00033.

7.103

(a) 0.37581.
(b) 0.0057.

7.105

(a) P(X > 592) = 0.69.
(b) 0.478.
(c) They are not independent events and the answer to (b) is incorrect.

8.1
(a) The sampling distribution of a statistic is the distribution of the values of the statistic in all possible samples of the same size from the same population. We often generate the empirical sampling distribution using some form of simulation.

8.3
The answer is (b).

8.5
(a) Histogram C because X takes on only two values, 0 and 1, with probabilities of 0.80 and 0.20.
(b) Histogram B because we know with a large sample size the distribution of the sample proportion will be approximately normal, centered at the true $p = 0.20$.

8.7
(a) Yes.
(b) Yes.
(c) No, since $np < 5$.

8.9
Note that $\hat{p} \approx N(0.50, 0.025)$ so $P(\hat{p} \leq 0.45) = P(Z \leq -2) = 0.0228$;
or normalcdf(-E99, 0.45, 0.5, 0.025) = 0.02275.

8.11
Answer is (c).

8.13
(a) Approximately normal with mean of 0.85 and standard deviation of 0.0357.
(b) The sample proportion is 90/100 or 0.90.
(c) p-value = $P(\hat{p} \geq 0.90) = P(Z \geq 1.40) = 0.0808$; or normalcdf(0.90, E99, 0.85, 0.0357) = 0.0807.
(d) Since the p-value is > 0.05, the new method is not significantly more accurate at a 5% significance level.

8.15
(a) The true standard deviations are given by:
$n = 50, p = 0.1$ standard deviation = 0.0424; $n = 50, p = 0.3$ standard deviation = 0.0648;
$n = 50, p = 0.7$ standard deviation = 0.0648;
$n = 50, p = 0.5$ standard deviation = 0.0707; $n = 100, p = 0.5$ standard deviation = 0.050.
(b) and (c) Answers will vary.

8.17
False. The central limit theorem states that for simple random samples of size n, where n is large, the distribution of $\overline{X}$ is approximately normal, it will not be exactly normal. The approximation is better for larger the sample size.

8.19

(a) Since X is $N(1250, 150)$, the probability is

$$P(1200 < X < 1400) = P\left(\frac{1200-1250}{150} < Z < \frac{1400-1250}{150}\right) = P(-.33 < Z < 1) = 0.4719$$

or with normalcdf(1200, 1400, 1250, 150) = 0.4719.

(b) Since $\overline{X}$ is $N(1250, 150/6)$, the probability is

$$P(1200 < \overline{X} < 1400) = P\left(\frac{1200-1250}{150/\sqrt{36}} < Z < \frac{1400-1250}{150/\sqrt{36}}\right) = P(-2 < Z < 6) = 0.9772$$

or with normalcdf(1200, 1400, 1250, 25) = 0.9772.

(c) The probability in part (b) is higher since the distribution of $\overline{X}$ is more concentrated around 1,250 (i.e. has a smaller standard deviation) than that of X.

8.21

(a) $P(X > 240) = P\left(Z > \dfrac{240-170}{30}\right) = P(Z > 2.33) = 0.0098$ or with normalcdf((240, E99,170, 30) = 0.0098.

(b) $\overline{X}$ is $N(170, 30/2)$ so the standard deviation is 15.

(c) $P(\overline{X} > 190) = P\left(Z > \dfrac{190-170}{15}\right) = P(Z > 1.33) = 0.09121$ or with normalcdf(190, E99, 170, 15) = 0.09121.

8.23

Fail to operate implies that the total weight exceeds 5000 or equivalently that the average or mean weight exceeds

555.6. The probability is $P(\overline{X} > 555.6) = P\left(Z > \dfrac{555.6-540}{45/\sqrt{9}}\right) = P(Z > 1.04) = 0.1492$

or using normalcdf(555.6, E99, 540, 15) = 0.1492.

8.25

(a) Histogram C.

(b) Histogram F.

8.27

(a) 0.61

(b) $E(X) = 1.2$

(c) $\overline{X}$ will have a $N(1.2,\ 0.9/10)$ distribution.

(d) $P(\overline{X} < 1) = $ normalcdf(-E99,1,1.2,.9/10) = 0.013134

8.29

Note that the total number of tickets desired being less than or equal to 220 is equivalent to having the sample mean number of tickets desired for 100 students being less than or equal 2.2. The sample mean has a $N(2.1, 0.2)$ distribution and the probability is given by: $P(\overline{X} \leq 2.2) = P(Z \leq 0.5) = 0.6915$ or use normalcdf(-E99, 2.2, 2.1, 0.2).

8.31

(a) False.

(b) True.

(c) False.

(d) False.

8.33

(a) The answer is Histogram B.

(b) No. If the variability remains the same, the sampling distribution does not depend on the population size.

8.35

The population proportion is $p = 0.60$.

So the sample proportion $\hat{p}$ has approximately a $N(0.6, \sqrt{.24/100}\,)$ distribution, that is, $\hat{p} \approx N(0.6, 0.049)$.

$P(\hat{p} < 0.5) = P(Z < (0.5 - 0.6)/\,0.049\,) = 0.0206$, or using normalcdf(-E99, 0.5, 0.6, 0.049) = 0.0206. Thus it is very unlikely to observe a $\hat{p} < 0.50$ for a random sample of size 100 if the true proportion is 0.60.

8.37

(a) The three remaining possible samples are: (9,1), (9,5), (9,9). The sample means for all nine possible samples are: 1, 3, 5, 3, 5, 7, 5, 7, 9.

(b) 3/9.

(c) 5.

8.39

(a) The mean is 4 and the variance is 8.

(b) Using TI with Seed = 83   Using random number table with
                                                         row 10, column 1

Number 1	Number 2	Mean	Number 1	Number 2	Mean
8	0	4	8	4	6
6	8	7	6	8	7
8	0	4	4	2	3
6	6	6	8	8	8
4	8	6	0	6	3
8	6	7	0	8	4
6	4	5	8	6	7
2	4	3	6	2	4
2	8	5	0	0	0
8	8	8	0	8	4
8	6	7	8	8	8
0	4	2	6	4	5
8	6	7	4	8	6
8	4	6	8	6	7
4	2	3	4	2	3
4	4	4	2	8	5
2	0	1	8	6	7
2	0	1	8	8	8
8	2	5	8	2	5
8	8	8	2	6	4
6	2	4	0	6	3
8	8	8	6	8	7
8	0	4	6	0	3
4	4	4	8	0	4
2	2	2	0	6	3

Sampling Distributions 55

(c) Here are the histograms of the sample means of part (b):

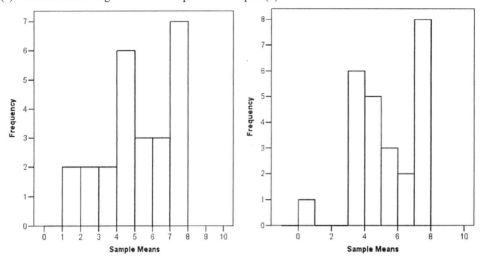

(d) Using TI with Seed = 124

Number 1	Number 2	Number 3	Number 4	Number 5	Mean
4	0	8	0	0	2.4
4	8	0	6	8	5.2
2	6	4	6	2	4.0
8	6	4	8	8	6.8
8	2	4	8	4	5.2
0	8	6	4	6	4.8
6	6	4	4	4	4.8
2	0	0	0	0	0.4
2	8	4	4	4	4.4
0	4	4	0	2	2.0
0	0	8	2	0	2.0
6	0	2	4	6	3.6
8	2	2	6	2	4.0
2	0	2	6	8	3.6
4	2	6	4	2	3.6
8	2	6	0	0	3.2
8	6	0	6	2	4.4
4	2	6	2	8	4.4
6	0	4	8	2	4.0
4	4	2	8	8	5.2
0	2	0	6	2	2.0
0	4	0	2	2	1.6
0	6	0	2	6	2.8
0	4	0	8	8	4.0
8	4	6	6	2	5.2

Using the random number table with row 20, column 1:

Number 1	Number 2	Number 3	Number 4	Number 5	Mean
0	0	6	6	2	2.8
8	8	0	8	4	5.6
2	6	8	0	6	4.4
6	6	8	8	6	6.8
0	2	8	4	6	4.0
0	4	8	8	6	5.2
0	2	6	2	8	3.6
0	4	4	8	6	4.4
6	2	4	8	8	5.6
2	8	4	4	6	4.8
0	2	0	6	8	3.2
0	2	2	0	4	1.6
4	2	2	8	0	3.2
4	2	6	6	6	4.8
6	0	4	2	8	4.0
2	4	6	4	8	4.8
4	2	2	2	4	2.8
2	4	0	4	0	2.0
0	2	0	6	6	2.8
4	6	8	2	6	5.2
8	4	8	0	6	5.2
6	4	6	2	2	4.0
4	2	2	6	0	2.8
4	4	2	2	6	3.6
2	6	0	2	4	2.8

(e) Here are the histograms of the sample means of part (d):

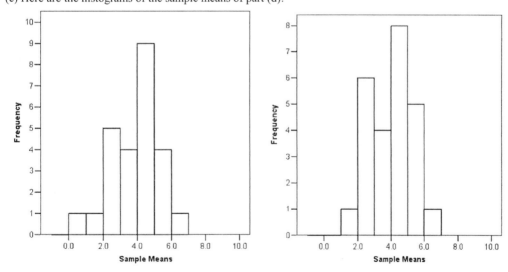

(f) The second plots have less variation about the mean. The data are more clustered around the true mean of 4.

8.41

(a) $1 - (0.39 + 0.24 + 0.14 + 0.09 + 0.05 + 0.03) = 0.06$.

(b) Histogram C. Since the sample size n is large, the distribution of the sample mean will be approximately normal and centered at the population mean.

8.43

(a) $P(X > 180) = P\left(Z > \dfrac{180 - 173}{30}\right) = P(Z > 0.2333) \approx 0.4078$

or use normalcdf(180,E99,173,30) = 0.4078.

(b) $P(\overline{X} > 180) = P\left(Z > \dfrac{180 - 173}{30 / \sqrt{36}}\right) = P(Z > 1.4) = 0.0808$.

or use normalcdf(180,E99,173,5) = 0.0808.

8.45

(a) $\overline{X} \sim N(100, {}^{10}\!/_{\sqrt{49}})$ so the standard deviation is 1.429.

$P(\overline{X} \le 95) = P\left(Z \le \dfrac{95 - 100}{1.429}\right) = P(Z \le -3.5) = 0.0002$

or normalcdf(-E99, 95 , 100, 1.429) ~ 0.00023.

(b) Find n such that $P\left(98 \le \overline{X} \le 102\right) = 0.95$

$P\left(\dfrac{98 - 100}{{}^{10}\!/_{\sqrt{n}}} \le \overline{X} \le \dfrac{102 - 100}{{}^{10}\!/_{\sqrt{n}}}\right) = 0.95$

So we have

Thus $\dfrac{102 - 100}{{}^{10}\!/_{\sqrt{n}}} = 1.96$ and $1.96({}^{10}\!/_{\sqrt{n}}) = 2$. So $\sqrt{n} = 1.96(10)/2 = 9.8$.

So n must be at least 97.

8.47

No, the variability in the possible values of the sample mean is smaller with a larger sample size. So the second organization with the larger sample size will not be likely to get a higher sample mean.

8.49

Pablo is correct - as the size of each sample used increases, the sampling distribution has less variability. Eduardo is also correct, the value of the parameter p does influence the variability for the sampling distribution of a sample proportion.

9.1

(a) (i) $H_0: p = 0.50$ versus $H_1: p > 0.50$.

   (ii) $H_0: p = \frac{1}{2}$ versus $H_1: p \neq \frac{1}{2}$.

(b) (i) A Type I error would be concluding that the proportion of democrats in Wayne County is more than 0.50, when it is not. A Type II error would be concluding that the proportion of democrats in Wayne County is not more than 0.50, when it is.

   (ii) A Type I error would be concluding that the proportion of male births in Wayne County is not ½, when it is. A Type II error would be concluding that the proportion of male births in Wayne County is ½, when it is not.

9.3

(d) It is equal to α.

9.5

(b) p-value = 0.005.

9.7

(a) The population of pregnant women who work with a computer 1-20 hours per week.

(b) The test statistic formula is $Z = \dfrac{\hat{p} - p_0}{\sqrt{\frac{p_0(1-p_0)}{n}}}$. The observed test statistic value is $z = 1.4772$. We cannot reject H_0 because the p-value = 0.0698 > 0.01.

(c) No, the results are not statistically significant at the level 0.01.

9.9

(a) $\hat{p} = 0.7289$.

(b) The test can be performed using the 1-PropZTest with the calculator. The test statistic is $z = 1.323$ and the corresponding p-value is $P(Z \geq 1.323) = 0.0930$. We cannot reject the null hypothesis.

9.11

$H_0: p = 0.2$ versus $H_1: p > 0.2$; $\hat{p} = 0.72$. The p-value = $P(Z \geq 9.19) = 0$. Reject H_0.

9.13

H_0: The proportion of all reports with discrepancies, $p = 0.20$
H_1: The proportion of all reports with discrepancies, $p < 0.20$
The test statistic value = -1.5811 and p-value = 0.0569. We cannot reject H_0 at the level 0.05. The results are not statistically significant at 0.05.

9.15

(a) It may be a sample of typical working mothers with young children.

(b) $H_0: p = 0.5$ versus $H_1: p \neq 0.50$.

(c) No. We do not know how many of the 1200 had children under six.

9.17

(a) The sample proportion is 0.45.

(b) Note that the standard deviation of the sample proportion under H_0 is given by $\sqrt{\dfrac{(0.2)(0.8)}{600}} = 0.0163$. So the p-value = normalcdf(0.45, E99, 0.20, 0.0163) which is nearly 0.

(c) Reject H_0.

(d) The power of the test is the probability of rejecting H_0 when H_1 is true. Since the significance level is 0.05, we need to find the critical value for when we would reject H_0. That is we need the 95th percentile for the H_0 model. This percentile can be found using the TI with invnorm(0.95, 0.2, 0.0163) = 0.2268. So the power of the test can be found as normalcdf(0.2268, E99, 0.40 , 0.02) which is nearly 1. Note that the standard deviation of the sample proportion under H_1 is given by $\sqrt{\dfrac{(0.4)(0.6)}{600}} = 0.02$.

9.19

(a) $\hat{p} = 17/40 = 0.425$

(b) The 99% confidence interval for the population proportion is (0.22367, 0.62633). It can be found from the 1-PropZInt with the calculator or using the confidence interval formula with $z^* = 2.576$.

(c) The margin of error = E = 0.20133.

9.21

(a) The 95% confidence interval for the population proportion is (0.66784, 0.73216)

(b) The population proportion p represents the true proportion of operating vehicles that are equipped with air bags.

9.23

(a) 633

(b) A statistic.

(c) $\hat{p} = 0.63$.

(d) Standard error of the sample proportion is: $\sqrt{\dfrac{(0.63)(0.37)}{1004}} = 0.0152$.

(e) The 95% confidence interval is (0.60062, 0.66033)

(f) The margin of Error = E= 0.02986 $\approx$ 0.03. Yes, it is close to the stated value.

9.25

The 92% confidence interval is (0.0151, 0.1099).

9.27

A confidence interval is constructed for estimating a population parameter, not a sample statistic.

9.29

(a) $\hat{p}$ is approximately N(p, $\sqrt{\dfrac{p(1-p)}{n}}$) where $n = 4000$.

(b) No, the sample proportion is an unbiased estimator of the population proportion.

(c) Yes, the sample size is in the denominator of the variance.

(d) The confidence interval of (0.51, 0.55) was made with a method, which if repeated, would result in many 95% confidence intervals. We would expect 95% of these possible intervals to contain the population proportion p.

9.31

(a) The interval provides plausible values for the true population proportion of abused children with above average anxiety level, based on this sample of 60 such children.

(b) If we repeat this procedure many times, we would expect the true proportion p to lie inside about 95% of the confidence intervals produced in this manner.

9.33

A sample of $n \geq (2.326/0.06)^2 = 1502.85$ is required. Thus a sample of at least 1503 students is needed.

9.35

(a) (i) – (iv) are false. (v) True.

(b) $n \geq (1.96/0.08)^2 = 601$

9.37

(a) A statistic.

(b) (0.79199, 0.92801)

(c) E = 0.06801.

(d) $n \geq (1.96/0.08)^2 = 601$.

(e) $H_0: p = 0.87$ versus $H_1: p < 0.87$.

(f) The test statistic value is $z = -0.2974$. The p-value = 0.3831. You cannot reject H_0 at the 0.05 level.

9.39

$H_0: p = 2/3$ versus $H_1: p \neq 2/3$.

9.41

(a) $H_0: p = 0.72$ versus $H_1: p > 0.72$. One-sided to the right.

(b) $H_0: p = 0.90$ versus $H_1: p < 0.90$. One-sided to the left.

(c) $H_0: p = 0.60$ versus $H_1: p \neq 0.60$. Two-sided.

9.43

(a) No, since we do not know the number of pregnant women.

(b) (0.1346, 0.1659).

(c) Approximately 100.

9.45

Since there are very few yes's in the sample, the Wilson estimate is recommended.

$$\widetilde{p} = \frac{3+2}{200+4} = 0.0245.$$

9.47

(a) Note that the number who have not put aside money in the sample was 52% of 1156 or 611. The 95% confidence interval is given by (0.4911, 0.5487).

(b) The 95% confidence level implies that if we were to repeat the sampling process a large number of times, then we'd expect approximately 95% of the intervals to contain the true population proportion p. The margin of error for the confidence interval in part (a) is 0.029 that is approximately 2.9%.

(c) The time of the day the calls were placed could lead to a bias in terms of who was home to respond.

(d) A multistage sampling plan was used.

9.49

(a) $H_0: p = 0.50$ versus $H_1: p > 0.50$.

(b) $z = 1.7215$; p-value = 0.0426.

(c) Yes, the p-value is smaller than 0.05.

9.51

(a) $H_0: p = 0.60$ versus $H_1: p > 0.60$.

(b) The test statistic value = 1.4434. The p-value = 0.0745. The results are not statistically significant. We cannot reject H_0 at the 0.01 level.

(c) No, we just need to have a sample size that is large enough so the distribution of the sample proportion is approximately normal.

9.53

(a) (0.44126, 0.51874).

(b) The margin of error = E = 0.03874

(c) The margin of error will increase to 0.06066.

9.55

(a) An observational study.

(b) No.

(c) No.

10.1
(a) H_0: $\mu = 33$ °F versus H_1: $\mu < 33$ °F.
(b) H_0: $\mu = 26$ years versus H_1: $\mu > 26$ years.
(c) H_0: $\mu = 200$ versus H_1: $\mu \neq 200$.

10.3
(a) H_0: $\mu = 5000$ versus H_1: $\mu < 5000$.
(b) $z = -2.5$.
(c) p-value = 0.0062.
(d) Yes, since the p-value < 0.01.

10.5
H_0: $\mu = 60$ versus H_1: $\mu < 60$, the test statistic $= z = -2.5$, p-value $= 0.0062$. We reject H_0.

10.7
(a)

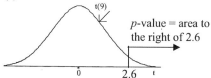

p-value = area to the right of 2.6

Using the calculator the p-value = tcdf(2.6, E99, 9) = 0.014369. Using the t-distribution table the bounds for the p-value are given by: $0.01 < p$-value < 0.025.
(b)

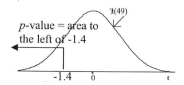

p-value = area to the left of -1.4

Using the calculator the p-value = tcdf(-E99, -1.4, 49) = 0.0839. Using the t-distribution table the bounds for the p-value are given by: $0.05 < p$-value < 0.10.
(c)

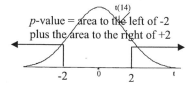

p-value = area to the left of -2 plus the area to the right of +2

Using the calculator the p-value = 2tcdf(-E99, -2, 14) = 2(0.03264) = 0.06529. Using the t-distribution table the bounds for the p-value are given by: $0.05 < p$-value < 0.10.

10.9

(a) H_0: $\mu = 16$ versus H_1: $\mu \neq 16$.
(b) Assume the distribution of the amount of syrup is normal.
(c) The company would turn off the equipment to investigate and lose money.
(d) The company either loses money or customers would be dissatisfied.
(e) Using the calculator with option 2:TTest we have: $t = -1.28$, p-value $= 0.2167$. Using the t-distribution table we have that the p-value is $0.2 < p$-value < 0.25 So we would not reject the null hypothesis.
(f) Our advice to the company is to let the machine continue to run.

10.11

(a) The histogram and QQ plot are given below and show no strong departures from the assumption of a normal model for the response.

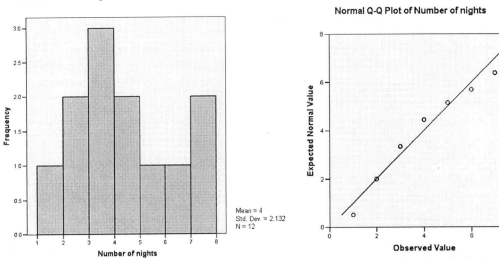

(b) Using the calculator with option 2:TTest we have $t = 1.46$; p-value $= 0.0858$. Using the t-distribution table we have that $0.05 < p$-value < 0.10.
(c) Since the p-value is less than 0.10, we reject the null hypothesis.
(d) Yes, a Type I error. The probability of having committed an error is either 0 or 1.

10.13

(a) H_0: $\mu = 12$ versus H_1: $\mu > 12$.
(b) Using the calculator with option 2:T-Test we have: $t = 1.3765$, p-value $= 0.1136$. Using the t-distribution table we have that $0.10 < p$-value < 0.20.
(c) Assume the data are a random sample and concentration has a normal distribution.
(d) If the mean concentration were 12 mg/kg, we would see a test statistic of 1.376 or even larger about 11.4% of the time in repeated sampling.
(e) Do not reject the null hypothesis since the p-value is $> \alpha$.

10.15

H_0: $\mu = 50$ versus H_1: $\mu < 50$.
Using the calculator with option 2:TTest we have: $t = -2.4$, p-value $= 0.0109$. Using the t-distribution table we have that $0.01 < p$-value < 0.05. So we reject H_0 at $\alpha = 0.05$.

10.17

(a) Using the calculator with option 2:TTest we have: $t = -.8.8$, p-value is nearly 0. Using the t-distribution table we have that p-value is much less than 0.005 (nearly 0).
(b) Yes, the t-test is robust against modest departures from normality.

10.19

(a) $\bar{x} = 13.5$.

(b) Using the calculator with option 8:TInterval we have: (13.094, 13.906). The value of t^* from the table would be the conservative 2.042 (using 30 degrees of freedom).

(c) If we repeat this method over and over, yielding many 95% confidence intervals for μ, we expect 95% of these intervals to contain the true parameter value.

10.21

(a) Using the calculator with option 8:TInterval we have: (434.99, 675.41). With the t-distribution table we would use $t^* = 2.093$.

(b) The margin of error = E = 120.21.

(c) If we repeat this method over and over, yielding many 95% confidence intervals for the true mean, we expect about 95% of these intervals to contain μ.

10.23

The sample mean is 49.2 and the margin of error is E = 5.

10.25

Only student 4's answer is correct.

10.27

(a) The standard deviation is s=2.82. The 90% confidence interval is given as:

$$10.44 \pm 1.7081\left(2.82\middle/\sqrt{26}\right) \Rightarrow 10.44 \pm 0.945 \Rightarrow [9.495, 11.385].$$

(b) If repeated samples were obtained (of the same size, from the same population), we would *expect* about 90% of the resulting confidence intervals generated to contain the population mean μ.

(c) The corresponding boxplot is shown below.

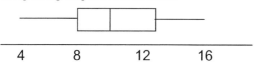

```
        4       8       12      16
```

10.29

(a) The sample proportion is the midpoint of the interval, given as 71%. So 71 out of the 100 completed their degree in 5 years or less.

(b) The only valid interpretation is (iii).

10.31

(a) Using the calculator with option 8:TInterval we have: (75.9,84.1). With the t-distribution table we would use $t^* = 1.740$.

(b) E = 4.1

(c) No, 82 is inside the confidence interval.

(d Narrower.

10.33

(a) Yes.

(b) Can't tell.

(c) Can't tell.

(d) No.

(e) Yes.

10.35

(a) The sample mean is the midpoint of the interval, namely, 15.9 ounces.

(b) True, False, False.

(c) Fail to reject H_0. The value of 16 is in the 90% confidence interval, and thus will also be in the wider 95% confidence interval for plausible values of the population mean.

(d) No. We are not too concerned since our sample size of $n = 50$ boxes is somewhat large. We will be able to rely on the central limit theorem for the approximate normality of the sample mean.

10.37

The correct answer is (b).

10.39

(a) $H_0: \mu = 17.1$ versus $H_1: \mu < 17.1$.

(b) Using the calculator with option 2:TTest we have: $t = -2.3159$, p-value $= 0.0229$. Using the t-distribution table we have: $0.01 < p$-value < 0.025.

(c) We reject the null hypothesis. It does appear that the new species has a faster germination time on average.

10.41

(a) $P(X \leq 285) = P(Z \leq (285 - 270)/10) = \text{normalcdf}(-E99, 285, 270, 10) = 0.93319$.

(b) (i) P(at least one among the four pregnancies lasts more than 285 days) =
$1 - P(\text{all pregnancies last less or equal to 285 days}) = 1 - 0.93319^4 = 0.2416$.

(ii) $P(\overline{X} < 285) = P(Z < (285 - 270)/(10/2) = \text{normalcdf}(-E99, 285, 270, 5) = 0.99865$.

(c) $H_0: \mu = 270$ versus $H_1: \mu < 270$.

(d) $t = -1.3333$, p-value $= 0.101858$.

(e) You cannot reject the null hypothesis at the level 0.05.

(f) Women who work outside the home do not appear to have a different mean pregnancy length than for the general population of women.

10.43

(a) Using the calculator with option 8:TInterval we have: (625.35,658.65). With the t-distribution table we would use $t^* = 1.753$.

(b) (i) False.

(ii) False.

10.45

(a) False.

(b) False.

(c) False.

(d) True.

(e) Fail to reject H_0.

10.47

(a) The apartments in the suburb.

(b) Living area of the apartments in square feet.

(c) 1325 square feet.

(d) Using the calculator with option 8:TInterval we have: (1316.7, 1333.3). With the t-distribution table we would use the formula with a value of $t^* = 2.000$ (using the conservative 60 degrees of freedom).

(e) No. The true mean is or is not inside the interval.

(f) Increase the confidence level or decrease the sample size.

10.49

(a) Using the calculator with option 7:ZInterval we have (11.397, 11.603). With the standard normal table we would use $z^* = 2.576$.

(b) 0.103.

10.51

(a) Min = 230; Q_1 = 250; Median = 265; Q_3 = 290; Max = 310.

(b) With the calculator you could use the following window:

Xmin =225
Xmax=325
Xscl=10
Ymin=-1
Ymax=6
Yscl=1
Xres=1
Histogram

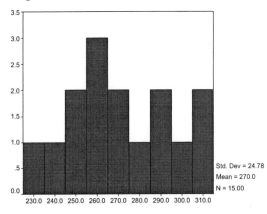

PRICE

(c) H_0: μ = 258 versus H_1: μ > 258.

(d) Using the calculator with option 2:T-Test we have: t =1.875; p-value = 0.0408. Using the t-distribution table we have: 1.761. We would reject the null hypothesis since 0.0408 < 0.05. The data are statistically significant.

10.53

Results will vary.

11.1
Paired - Each female twin is directly linked to her twin brother.

11.3
Paired - Each woman is directly related (linked) to her husband.

11.5
(a) The male students can be paired up in the following pairs: Ronald & Lee and Kyle & Pablo; Ronald & Kyle and Lee & Pablo; Ronald & Pablo and Lee & Kyle. The female students can be paired up in the following pairs: Kerri & Emily and Monica & Sonya, Kerri & Monica and Emily & Sonya; Kerri & Sonya and Emily & Monica. One of the 3 possible pairings for males can be combined with any of the 3 possible pairings for females, resulting in nine possible pairings. One example is Ronald & Kyle, Kerri & Sonya, Lee & Pablo, and Emily & Monica.

(b) Ronald & Kyle (20), Kerri & Sonya (18), and then one of the three possible pairs for 19 year olds: Emily & Lee with Pablo & Monica, Emily & Pablo with Lee & Monica, or Emily & Monica with Lee & Pablo.

(c) One possible assignment: Let Ronald = 1, Lee = 2, Kyle = 3, Pablo = 4. Also let Kerri = 1, Emily = 2, Monica = 3, Sonya = 4. One possible method: Step 1: Select 2 random numbers between 1 and 4. These two will form the first male pair. The remaining two form the second male pair. Step 2: Repeat Step 1 for the females. Using the TI: Step 1 (seed = 18): 2, 2 (skip), 1. Male pairs: Ronald & Lee, Kyle & Pablo. Step 2 (seed = 33): 3, 1. Female pairs: Kerri & Monica, Emily & Sonya. Using the random number table: Step 1 (row 8, column 1): 3, 1. Male pairs: Ronald & Kyle, Lee & Pablo. Step 2 (row 22, column 6): 4, 2. Female pairs: Emily & Sonya, Kerri & Monica.

(d) One possible assignment: Let Emily = 1, Lee = 2, Pablo = 3, Monica = 4. One possible method: Step 1: The two 20-year students automatically form pair 1: Ronald & Kyle. The two 18-year old students automatically form pair 2: Kerri & Sonya. Step 2: Select 2 random numbers between 1 and 4. These two will form the first 19-year old pair. The remaining two form the second 19-year old pair. Using the TI: Step 2 (seed = 28): 4, 4 (skip), 2. 19-year old pairs: Lee & Monica, Emily & Pablo. Using the random number table: Step 2 (row 10, column 1): 4, 3. 19-year old pairs: Pablo & Monica, Emily & Lee.

11.7
(a) You might take a random sample of direct admits and record the waiting times and then take an independent random sample of ER admits and record the waiting times. You might instead introduce pairing by recording waiting times for ER and direct admits, matching them by the day or time of day; or perhaps matching them by type of injury or situation.

(b) A paired design would be most reasonable here. Select a representative set of groceries and measure the prices of these same items at both stores. We would be interested in looking at the difference in prices for each grocery item.

(c) Independent random samples of male versus female students could be obtained and the response, amount of time spent studying (over some specified time frame) could be recorded. Or you may elect to do a paired design by matching up students by their major or school (business, law, etc.).

11.9
(a) The sample mean difference is 4.2 pounds.
(b) The sample standard deviation of the differences is 6.3 pounds.
(c) The standard error of the mean is $\dfrac{6.3}{\sqrt{12}} = 1.82$ pounds.

11.11
(a) $H_1 : \mu_D > 0$

(b) The data are a random sample from a normal population with unknown population standard deviation. The study background states that the 14 children were selected at random. The Q-Q plot below shows no strong departures from the normality assumption.

Normal Q-Q Plot of New - Regular

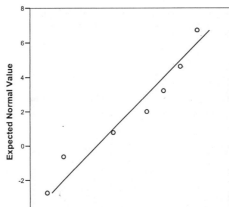

(c) From the table we find the sample mean of the differences is 2 and the sample standard deviation of the differences is 3.464. So the observed test statistic is $t = \dfrac{2}{\dfrac{3.464}{\sqrt{7}}} = 1.5275$. The p-value is

$P(T \geq 1.5275) = 0.0887$, where T has n-1 = 6 degrees of freedom. We fail to reject H_0 at the 5% level and conclude that the new program does not seem to improve creative thinking. With the TI we would use the 2: TTest option.

(d) The correct answer is (i).

11.13

(a) Let $\mu_D = \mu_1 - \mu_2$. The hypotheses are: $H_0 : \mu_D = 0$ vs. $H_1 : \mu_D \neq 0$.

(b) The data are a random sample from a normal population with unknown population standard deviation.

(c) There is only one pair (the third pair) for which the highest weight is on scale 2.

(d) Excluding the third pair, we find from the table we find the sample mean of the differences is 2.6 and the sample standard deviation of the differences is: 2.302. The critical value we need is $t^* = 2.776$. So the 95% confidence interval is given by: $2.6 \pm 2.7764 \left(\dfrac{2.6}{\sqrt{5}} \right)$, or (-0.258, 5.458). With the TI we would use the 8: TInterval option.

(e) The answer is (ii). The p-value would be greater than 0.05 since 0 is included in the interval, so we fail to reject the null hypothesis.

11.15

(a) The observed sample mean is 1.45 standard errors above the hypothesized mean of 0.

(b) The p-value would be 0.190/2 = 0.095.

(c) No, we would not reject the null hypothesis because the p-value is greater than α.

(d) No, because the alternative hypothesis is one-sided and a confidence interval can only be used for a two-sided test.

11.17

(a) $H_0 : \mu_D = 0$ versus $H_1 : \mu_D > 0$ where D = second score – first score.

(b) This is a paired t-test so we need to assume that the differences are a random sample from a normal population.

(c) The standard error of the mean difference $= s_D \big/ \sqrt{n} = 40.4 \big/ \sqrt{4} = 20.2$. We estimate the average distance of the possible sample mean differences from μ_D to be about 20.2 points.

(d) We have $\bar{d} = 100/4 = 25$. So the observed test statistic is $t = 25/20.2 = 1.237$. Using Table IV: $0.15 < p$-value < 0.20. Using the TI: p-value $= 0.152$. Decision: fail to reject H_0. We conclude that there is not sufficient evidence to say students improve their score the second time they take the SAT.

(e) (ii) ... about 5% of the time.

11.19

(a) The differences are: 7, 6, -1, 5, 6, 1

(b) $\bar{d} = 4$

(c) $H_0 : \mu_D = 0$ versus $H_1 : \mu_D > 0$ where D = Before − After.

(d) The population of differences is assumed to be normal.

(e) $t = \dfrac{\bar{d} - 0}{s_D \big/ \sqrt{n}} = \dfrac{4 - 0}{3.225 \big/ \sqrt{6}} = 3.038$

(f) Using Table IV: $0.01 < p$-value < 0.02. Using the TI: p-value $= 0.0144$.

(g) Yes

(h) Type I Error

11.21

(a) Common (or equal) population standard deviations.

(b) Yes, the sample standard deviations are similar.

(c) We have $s_p = \sqrt{\dfrac{(16-1)(90)^2 + (16-1)(100)^2}{16+16-2}} = 95.1315$, so the 90% confidence interval is given by

$(600 - 550) \pm (1.697)(95.1315)\sqrt{\tfrac{1}{16} + \tfrac{1}{16}}$ => (-7.08, 107.08). Using the 0:2-SampTInt option on the TI would yield: (-7.086, 107.09).

11.23

(a) The sample means are 2.173 and 2.523. The sample standard deviations are 0.375 and 0.365. The pooled sample standard deviation is $s_P = \sqrt{\dfrac{(14)(0.375)^2 + (12)(0.365)^2}{26}} \approx 0.370$ and the 95% confidence interval for $\mu_1 - \mu_2$ is given by $(2.173 - 2.523) \pm (2.056)(0.370)\sqrt{\dfrac{1}{15} + \dfrac{1}{13}}$, or

(-0.6385, -0.0615). With a TI, you would use the 0:2-SampTInt option.

(b) The interval provides a range of plausible values for $\mu_1 - \mu_2$ at a 95% confidence level. If we repeat this procedure over and over, yielding many 95% confidence intervals for $\mu_1 - \mu_2$, we would expect that approximately 95% of these intervals would contain the true parameter value $\mu_1 - \mu_2$.

(c) You could test the hypotheses $H_0 : \mu_1 = \mu_2$ versus $H_1 : \mu_1 \neq \mu_2$. Since the confidence interval does not include the value of 0, we would reject the null hypothesis.

11.25

(a) The common population standard deviation is estimated by: $s_p = 2.1245$.

(b) The 90% confidence interval is given by: (2.3185, 5.6815)

(c) False.

(d) Since the 90% confidence interval does not contain the value of 0, the answer is yes.

11.27

(a) The hypotheses are: $H_0 : \mu_1 = \mu_2$ versus $H_1 : \mu_1 \neq \mu_2$.

(b) Strain 1: sample mean is 37.43, sample standard deviation is 3.69.
Strain 2: sample mean is 42.5, sample standard deviation is 3.89.

The pooled sample standard deviation is: $s_P = \sqrt{\dfrac{(6)(3.69)^2 + (7)(3.89)^2}{13}} \approx 3.80$.

The observed test statistic is: $t = \dfrac{(37.43 - 42.5)}{3.80\sqrt{\dfrac{1}{7} + \dfrac{1}{8}}} = $ -2.579.

The p-value is $2P(T \leq -2.579) = 0.0229$, where T has 13 degrees of freedom. With a TI, you would use the 0:2-SampTTest option.

(c) Since the p-value is less than 0.05, we reject the null hypothesis and conclude that the average weight for the two strains appear to be different.

(d) (i) Approximately 95% of the intervals produced with this method are expected to contain $\mu_1 - \mu_2$. (ii) We would fail to reject the null hypothesis since the value of -9 is in the confidence interval.

11.29

(a) $H_0 : \mu_1 = \mu_2$ versus $H_1 : \mu_1 > \mu_2$,
where 1 = Students with a C or higher and 2 = Students with D or E.

(b) We have $\bar{x}_1 = 4.6$, $\bar{x}_2 = 2.25$, $s_1 = 3.4$, $s_2 = 1.3$, $s_p = 2.7$, $t = 1.841$, and a p-value $= 0.0421$. With a TI, you would use the 0:2-SampTTest option. We would reject H_0 and conclude that the Students who earn a C or higher do appear to spend more hours per week outside of class on course work on average as compared to Students who receive a D or E.

(c) Each sample is a random sample from a normal population. The two population standard deviations are assumed to be equal. The two samples are assumed to be independent. The assumption of equal population standard deviation is somewhat suspect, based on boxplots or comparing the sample standard deviations. However, the sample sizes are nearly the same. If we do not assume equal population standard deviations, the test statistic would be t = 2.01 and the p-value would be 0.034, so the same conclusion would be reached. Normality seems reasonable.

11.31

(a) $H_0 : \mu_1 = \mu_2$ versus $H_1 : \mu_1 > \mu_2$

(b) $t = 1.839$, p-value $= \dfrac{0.073}{2} = 0.035$; Reject H_0 and conclude that the Strat method does appear to produce higher scores on average as compared to the Basal method.

(c) Same as in part (b).

(d) Similar. Note the two sample standard deviations are very close.

11.33

The original X data are: 1, 2, 3, 4, which have a mean of 2.5 and a variance of 1.25. The transformed Y data are: 7, 9, 11, 13, which have a mean of 10 and a variance of 5. Note that the mean of $Y = 2(\text{mean of } X) + 5 = 2(2.5) + 5 = 10$, and that the variance of $Y = (2)^2 \text{Var}(X) = 4(1.25) = 5$.

11.35

(a) The hypotheses are: $H_0 : p_1 = p_2$ versus $H_1 : p_1 > p_2$, where population 1 is the responses for patients who use carbolic acid and population 2 is the responses for patients who did not use the acid.

(b) The sample proportions are: $\hat{p}_1 = 34/40 = 0.85$ and $\hat{p}_2 = 19/35 = 0.543$. The pooled estimate of p is given by
$\hat{p} = \dfrac{34 + 19}{40 + 35} = 0.7067$.

(c) The observed test statistic is $z = \dfrac{(0.85 - 0.543)}{\sqrt{(0.707)(0.293)\left(\dfrac{1}{40} + \dfrac{1}{35}\right)}} = 2.915$ with a p-value of 0.0018. Since the

p-value is less than 0.05, we reject the null hypothesis and conclude that the presence or absence of carbolic acid does constitute a significant effect. Specifically, patients that use the acid have a better chance of recovery.

11.37

$n_1 = n_2 = \frac{1}{2}\left(\frac{1.96}{0.01}\right)^2 \Rightarrow 19{,}208$

11.39
Answers will vary.

11.41
(a) An observational study since no treatment was actively imposed.
(b) Population 1 = Parents of infants who died of SIDS (in Southern California between January 1989 and December 1992); Population 2 = Parents of healthy infants (in the same area and for the same time period).
(c) There is not much detail given. The sample size from each group was 200. It did state that parents of 200 *similar* healthy infants were selected, indicating a possibility of a paired design.

11.43
(a) The sampled differences represent a random sample from a normal population.
(b) The p-value for a one-sided, to the right, alternative hypothesis is $0.06/2 = 0.03$. Since the p-value is less than 0.05, we would reject the null hypothesis. Yes, it appears that the bonus plan did significantly increase sales, on average.

11.45
(a) The 22.3 miles per gallon is the sample mean, because we have a sample of ten old engines, and 22.3 is just the mean of that sample.
(b) In scenario 1, two unrelated samples of units are measured. In scenario 2, related or paired samples are measured.
(c) It would be easier to assess in scenario 2. Differences in the means between the two samples will better reflect differences between old and new engines and not also differences in driving habits between different drivers.
(d) First rewrite the hypotheses: $H_0 : \mu_1 - \mu_2 = 0$ versus $H_1 : \mu_1 - \mu_2 < 0$. The sample means are 22.3 and 23.1. The sample standard deviations are 2.003 and 3.178. The pooled sample standard deviation is

$s_P = \sqrt{\dfrac{(n_1 - 1)s_1^2 + (n_2 - 1)s_2^2}{n_1 + n_2 - 2}} \approx 2.65623$ and the observed test statistic is $t = \dfrac{(22.3 - 23.1)}{2.66\sqrt{\dfrac{1}{10} + \dfrac{1}{10}}} = \text{-}0.673$.

Based on the t-distribution with 9 degrees of freedom, the p-value is 0.2547. Comparing it to $\alpha = 0.05$, we will fail to reject the null hypothesis. At a 5% significance level, the new engines appear to consume as much gas as old engines, on average. We assume we have two independent random samples from normal populations with equal population variance.

(e) The 99% confidence interval for μ_D is given by: $1 \pm (3.2498)\left(\dfrac{2.16}{\sqrt{10}}\right)$, or (-1.22, 3.22). This interval provides a

range of plausible values for μ_D at a 99% confidence level. We expect 99% of such intervals to contain μ_D in repeated sampling. With the TI we would use the 7: TInterval option.

11.47
(a) The sample proportions are: Treatment group $\hat{p}_1 = 56/2051 = 0.0273$ and Placebo group $\hat{p}_2 = 84/2030 = 0.04137$.

(b) The hypotheses are: $H_0 : p_1 = p_2$ versus $H_1 : p_1 \neq p_2$.

The pooled estimate of p is given by $\hat{p} = \dfrac{56 + 84}{2051 + 2030} = 0.0343.$

The observed test statistic is $z = \dfrac{(0.0273 - 0.04137)}{\sqrt{(0.0343)(0.9657)\left(\dfrac{1}{2051} + \dfrac{1}{2030}\right)}} = \text{-}2.47.$

The two-sided p-value of 0.0135 for assessing if there is a difference is greater than 0.01. Thus we fail to reject the null hypothesis. It appears that there is not a significant difference in the incidence rates of cardiac events for the two groups. With the TI we would use the 6: 2-PropZTest option.

(c) We have (0.0273 - 0.04137)/0.04137 = -0.34, for a 34% reduction.

11.49

(a) 9.74

(b) $H_0 : \mu_D = 0$ (where D = Male – Female) and $H_1 : \mu_D > 0$

(c) $t = 2.216$; p-value = 0.030/2 = 0.015; Reject H_0 and conclude that the male sports do seem to dominate for private schools. With the TI we would use the 2: TTest option.

11.51

(a) H_0: $\mu_1 - \mu_2 = 0$ or $(\mu_1 = \mu_2)$ versus H_1: $\mu_1 - \mu_2 > 0$ or $(\mu_1 > \mu_2)$

(b) $s_p = \sqrt{10,039,260.33} = 3168.48$

(c) $t = 1.748$; p-value = 0.097/2 = 0.0485; Reject H_0. With the TI we would use the 4: 2-SampTTest option.

(d) It appears that they do.

(e) If we repeated this process many times, we would expect 95% of the confidence intervals generated to contain the true difference in the population means, $\mu_1 - \mu_2$.

(f) $t = 1.748$; p-value = 0.049; reject H_0. The conclusions are essentially the same.

11.53

(a) Since the degrees of freedom are $n - 1 = 20$, there were 21 pairs of participants and thus 21*2 = 42 women in the study.

(b) H_0: $\mu_D = 0$ versus H_1: $\mu_D < 0$ where D = Difference = menopause – regular cycle.

(c) p-value: 0.195/2 = 0.0975 Decision: Fail to Reject H_0.

(d) Type II error.

(e) The difference in total sleep time is normally distributed.

11.55

(a) Since we have two groups with different sample sizes of 25 and 23, it is not paired, but rather an independent samples design.

(b) This would be a paired design, taking all $n = 25 + 23$ children and recording two measurements for each child.

11.57

(a) Output I since this is a paired design.

(b) The value of the t-statistic equals 4.332. This implies that the observed mean fell 4.332 standard errors above the hypothesized value of 0.

(c) p-value = 0.000268/2 = 0.000134.

(d) Positive

12.1

We test the equality of the means by comparing two estimates of the common population variance σ^2. Both will be unbiased estimators if the means of the populations are equal. If the null hypothesis is not true, then the value of MSB/MSW, i.e., the F-ratio tends to be inflated, because the MSB overestimates σ^2.

12.3

(a) 2.74
(b) 2.99
(c) 5.64
(d) 7.95
(e) 2.33
(f) 1.93

12.5

(a) The answer is part (iii).
(b) $F(4, 35)$.

12.7

(a) Under H_0, the three populations all have the same normal distribution, with the same common mean and the same common population standard deviation. Under H_1, the three populations all have a normal distribution with at least one mean different from the others, but the standard deviations are the same.
(b) Under H_0, the F-test statistic has an $F(2,30)$ distribution.

12.9

(a) Since $I - 1 = 3$, there were 4 methods
(b) Since $n - I = 60$, there were $n = 64$ observations
(c) Using Table V, the p-value is less than 0.01. Using the TI we have p-value = Fcdf(25.4, E99, 3, 60) = 0.000000000099. The p-value is much less than 0.01, so the result is statistically significant.

12.11

(a) Experiment
(b) Number of Hours without Sleep (or Sleep Deprivation Level).
(c) Number failures (to detect a moving object).
(d) Using the TI: Group 1 = 9, 20, 4, 12, 13; Group 2 = 3, 15, 16, 2, 6; Group 3 = 17, 1, 7, 11, 10; leaving the remaining in Group 4 = 5, 8, 14, 18, 19.
      Using the table we let subject 1 have labels 01, 21, 41, 61, 81 and subject 2 have labels 02, 22, 42, 62, 82, and so on through subject 20 with labels 20, 40, 60, 80, 00. Group 1 = 5, 7, 13, 8, 17; Group 2 = 14, 19, 6, 15, 18; Group 3 = 16, 3, 20, 1, 11; leaving the remaining in Group 4 = 2, 4, 9, 10, 12.
(e) H_0: $\mu_1 = \mu_2 = \mu_3 = \mu_4$ versus H_1: at least one population mean μ_i is different.

12.13

(a) H_0: $\mu_1 = \mu_2 = \mu_3$ versus H_1: at least one population mean μ_i is different. Based on the output, the p-value = 0.012 < 0.05, so we reject H_0. There is evidence at the 5% significance level to indicate differences among the mean productions of the three lines.
(b) 50% of the time, Line #1 produced more than 282 or 283 units.
(c) The three independent random samples should be obtained from three normally distributed populations with same population standard deviations. By the side-by-side boxplots, the equal population standard deviations assumption seems reasonable. Provided the other assumptions hold, the test appears to be valid.

12.15

(a) The Anova table is presented below.

Source	SS	DF	MS	F
Between	104,855	3	34951.6	9.915
Within	70,500	20	3525	
Total	175,355	23		

(b) The estimate of the common population standard deviation is $\sqrt{MSW} = 59.37$.

(c) H_0: $\mu_1 = \mu_2 = \mu_3 = \mu_4$ versus H_1: at least one population mean μ_i is different.

(d) Using Table V: the p-value is less than 0.01. Using the TI: p-value = Fcdf(9.915, E99, 3, 20) = 0.000325.

(e) Since the p-value is smaller than α, we reject H_0. It appears that not all mean physical fitness scores are the same.

12.17

(a) At the 10% level we fail to reject H_0 since the p-value = 0.117 is larger than $\alpha = 0.10$.

(b) $\sqrt{MSW} = 1.80$.

(c) A large F-ratio means that MSB is overestimating the common population standard deviation. This happens when the population means are not all the same.

(d) The effect size is 14.970/(14.970+97.273) = 0.133.

12.19

H_0: $\mu_1 = \mu_2 = \mu_3$ versus H_1: at least one population mean μ_i is different, where 1 = Group A, 2 = Group B, and 3 = Group C.

Descriptives

MINUTES

	N	Mean	Std. Deviation
Group A	6	17.000	1.789
Group B	6	14.000	1.414
Group C	6	12.667	1.751
Total	18	14.556	2.431

ANOVA

MINUTES

	Sum of Squares	df	Mean Square	F	Sig.
Between Groups	59.111	2	29.556	10.726	.001
Within Groups	41.333	15	2.756		
Total	100.444	17			

Since the p-value is less than 0.01, we would reject H_0 and conclude that at least one population mean time is different. The three diets do not appear to be the same with respect to the mean time.

95% confidence interval estimates:

Group A: $17 \pm 2.13(\frac{1.66}{\sqrt{6}}) \rightarrow$ (15.56, 18.44).

Group B: $14 \pm 2.13(\frac{1.66}{\sqrt{6}}) \rightarrow$ (12.56, 15.44).

Group C: $12.667 \pm 2.13(\frac{1.66}{\sqrt{6}}) \rightarrow$ (11.22, 14.11).

12.21

Overall Type I error $= 1 - (1 - \alpha)^c$

c	P(Type I Error)
5	.226
10	.401
25	.723
50	.923
100	.994

12.23

The correct answer is (c).

12.25

(a) H_0: $\mu_1 = \mu_2 = \mu_3$ versus H_1: at least one population mean μ_i is different.

Source	SS	DF	MS	F
Between	223.96	2	111.98	31.11
Within	431.96	120	3.6	
Total	655.92	122		

p-value $= P(F \geq 31.11) \approx 0.000$, where F has $(2,120)$ degrees of freedom. With this very small p-value, we would reject H_0 and conclude that the mean time on market for at least one county is different from the others.

(b) We conclude that there is a significant difference between the mean time on market for County 1 and each of the other two counties (County 2 and County 3).

12.27

(a) The completed ANOVA table is as shown.

	SS	df	Mean Square	F	Sig.
Between Groups	2668.8	2	1334.4	15.77	.0003
Within Groups	1184.1	14	84.6		
Total	3852.9	16			

(b) 9.198.

(c) No, since the p-value is only 0.0003, much less than 0.05, we would reject the null hypothesis.

(d) Short versus medium; and short versus long.

(e) $39.2 \pm 2.98 \left(9.198 / \sqrt{6} \right) \rightarrow (28.01, 50.39)$.

12.29

		Factor A	
		Level 1	Level 2
Factor B	Level 1	20	30
	Level 2	50	60

12.31

(a)

Fertilizer	Water		
	A	B	C
I	28	31.67	23.33
II	33.67	31	25.67
III	28.67	35.67	29.33
IV	31	31.33	26.67

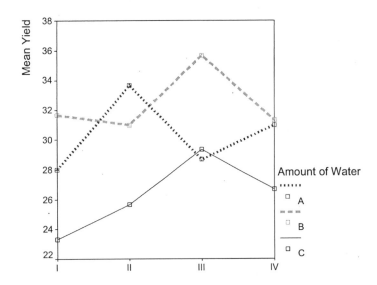

Fertilizer Type

(b) Since the profiles for the three amounts of water are not parallel, there is evidence of interaction. Overall water amount C performs the worst. Amount A works best with Fertilizer II and Amount B works best with Fertilizer III.

12.33

(a) The significance test for interaction is significant with a p-value of 0.006. The significance tests for the two main effects are also significant with p-values less than 0.0005.

(b) Although the profiles are not parallel, they do not cross. The largest difference between the two types of filters appears to occur for medium sized cars, while there is not much of a difference between the filter types for small or large sized cars.

12.35

(a) We fail to reject H_0: $\mu_1 = \mu_2 = \mu_3$ since the p-value of 0.117 is greater than α of 0.10.

(b) $\sqrt{MSW} = 1.801$.

12.37

(a) $H_0: \mu_1 = \mu_2 = \mu_3$ versus H_1: at least one population mean μ_i is different.

(b)

Source	SS	DF	MS	F
Between	89.6	2	44.8	5.1
Within	351.5	40	8.79	
Total	441.1	42		

(c) Since the MSW = 8.79, the estimate of the standard deviation 2.965.

(d) Using Table V, the p-value is between 0.01 and 0.05.

Using the TI, the p-value is Fcdf(5.1, E99, 2, 40) = 0.01064.

12.39

(a) $H_0: \mu_1 = \mu_2 = \mu_3$ versus H_1: at least one population mean μ_i is different.

(b) Since the MSW = 1393.2, the estimate of the standard deviation 37.33.

(c) No, the p-value of 0.0001 is less than 0.05.

(d) Answer (iii) is correct.

(e) $F(2, 12)$ distribution.

12.41

(a) $H_0: \mu_1 = \mu_2 = \mu_3$ versus H_1: at least one population mean μ_i is different

(b)

Level	Sample size	Sample mean	Sample standard deviation
Bottom	3	55.83	2.259
Middle	5	77.60	3.264
Top	5	52.70	2.255

The overall mean is 63.0 and we have $n = 13$ and $I = 3$.

$$SSB = 3(55.83 - 63)^2 + 5(77.6 - 63)^2 + 5(52.7 - 63)^2 = 1750.477$$

$$SSW = 2(2.259)^2 + 4(3.264)^2 + 4(2.255)^2 = 73.161$$

$$MSB = \frac{SSB}{I-1} = \frac{1750.4767}{3-1} = 875.238, \quad MSW = \frac{SSW}{n-I} = \frac{73.161}{13-3} = 7.316$$

$$F = \frac{MSB}{MSW} = \frac{875.238}{7.316} = 119.63 . \text{ Since the } p\text{-value is practically 0, we reject } H_0.$$

(c) The 3 populations follow a normal distribution with equal population standard deviations. We also need to have a simple random sample from each population. Finally, the 3 random samples are to be mutually independent.

13.1

(a) $\hat{y} = -30{,}000 + 7{,}000(30) = 180{,}000$.

(b) $\hat{y} = -30{,}000 + 7{,}000(50) = 320{,}000$. The estimation may not be meaningful since 5000 is outside of the range of *x* values used to calculate the regression equation.

13.3

(a) See scatterplot below. The scatterplot shows a positive linear relationship.

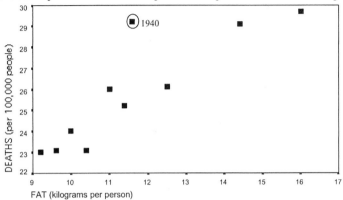

(b) $\hat{y} = 13.487 + 1.0649\,x$.

(c) $\hat{y} = 13.487 + 1.0649(13) = 27.33$ deaths per 100,000 people.

13.5

(a) A positive approximately linear relationship.

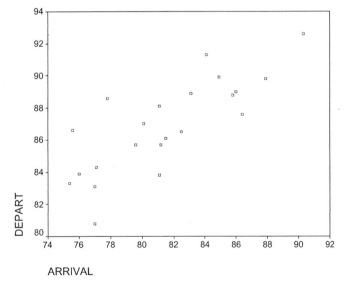

(b) $\hat{y} = 42.989 + 0.539\,x$.

13.7

(a) From the scatterplot shown below, a linear model does not seem to be appropriate.

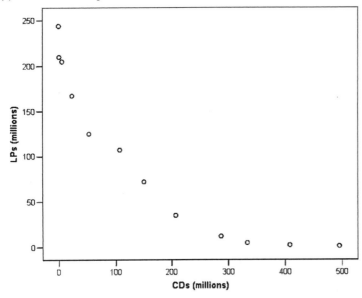

(b) This scatterplot shows a relationship that is approximately linear.

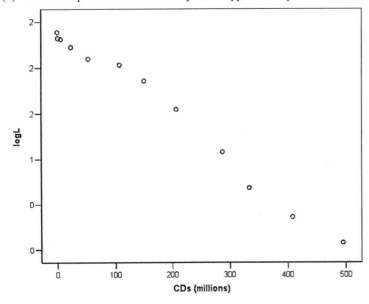

Predicted Log L = -0.004744(x) + 2.39929

(c) $L = 10^{(-0.004744(x) + 2.39929)}$

$= (10^{(-0.004744)})^x [10^{2.39929}]$

$= 0.98913598^x [250.778]$ or approximately $250.78(0.989)^x$

13.9

Above, Below, Below, Above, Below, On.

13.11

(a) Negative. The slope of the regression equation $b = -10.877$.

(b) The regression equation is $\hat{y} = 137.083 - 10.877\,x$.

The predicted number of grievances is $\hat{y} = 137.083 - 10.877(8) = 50.067$.

(c) Yes. The p-value $= 0.001 < 0.05$. The data are statistically significant at the 0.05 level.

13.13

(c) $r = -0.77$.

13.15

(c) As one variable increases, the other variable tends to decrease.

13.17 Answers will vary.

(a)

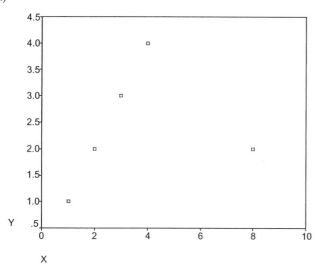

(b)

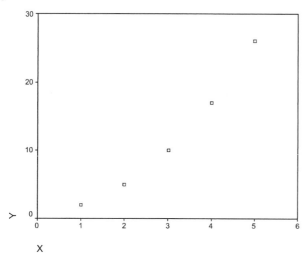

(c)

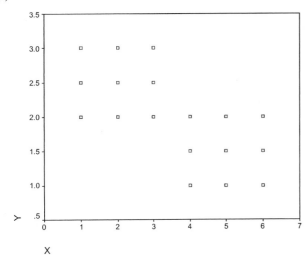

X

13.19

(a) 1

(b) 0

13.21

(a) $r = 0.908$.

(b) $\hat{y} = 6.863 + 0.862(30) = 32.72$.

(c) Unchanged.

13.23

(a) $r = -1$.

(b) There is no association because no matter what value of x we choose, the y value is always 3.

(c) $r = -1$.

(d) The data is not linear. Do not calculate r.

13.25

(a) Silence interval before response.

(b) Number of words used in the response.

(c) $\hat{y} = 70.837 - 0.282\,x$

(d) $r = -0.194$.

(e) Stay the same.

(f) Since the p-value=0.646 > 0.10, we cannot reject the null hypotheses. There is no significant (non-zero) linear relationship between the two variables.

13.27

(a)

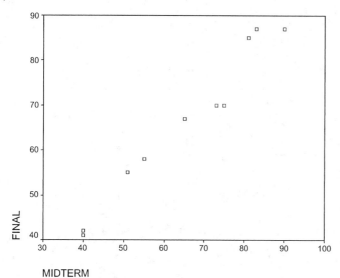

MIDTERM

(b) $\hat{y} = 4.8595 + 0.9394\,x$.

(c) $\hat{y} = 4.8595 + 0.9394(70) = 70.6175$.

(d) residual $= e = 70 - 73.436 = -3.436$.

13.29

(a) Scatterplot

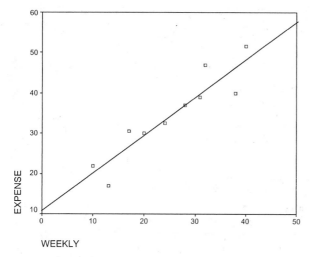

WEEKLY

(b) $\hat{y} = 10.528 + 0.953\,x$.

(c) Regression line is sketched in plot in part (a).

(d) $\hat{y} = 10.528 + 0.953(21) = \$3,054$ in annual maintenance expenses.

(e) $r = 0.9252$.

Relationships between Quantitative Variables 85

(f) Residual Plot

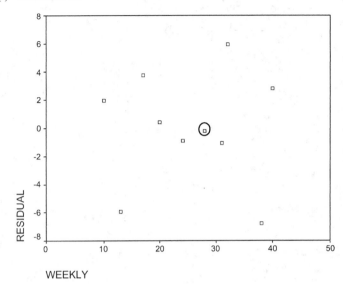

WEEKLY

(g) Yes, the residual plot shows a random scatter in a horizontal band around zero.

13.31
(a)

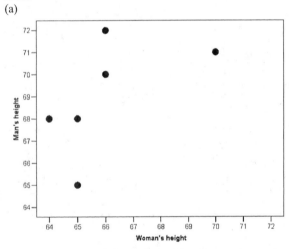

(b) Expect the correlation to be positive, but not close to 1.
(c) $r = 0.565$.
(d) No, a truism of statistics is "correlation, not causation".
(e) Unchanged.
(f) $r = 1$.
(g) 0.682
(h) (i) Unchanged.
     (ii) 1.73

13.33
This may be a consequence of the regression effect. The volunteers will probably have higher blood pressure on average than normal people.

13.35

(a) $r = 0.967$. R Square $= 0.936$ means that 93.6% of the total variation in vein blood flow rate can be explained by the linear regression between sphere blood flow rate and vein blood flow rate.

(b) $\hat{y} = 1.031 + 0.902\, x$.

(c) $[0.902 \pm 2.306(0.083)] = [0.902 \pm 0.1914] = (0.711\,,\,1.093)$.

(d) $H_0\colon \beta = 1$ versus $H_1\colon \beta \neq 1$. Since 1 is inside the confidence interval we cannot reject H_0 at the 0.05 level.

(e) $\hat{y} = 1.031 + 0.902\,(12) = 11.855$.

(f) $[11.855 \pm 2.306\,(1.7566)\sqrt{1 + 1/10 + \dfrac{(12 - 13.07)^2}{443.2}}\,] \Rightarrow (7.602,\,16.108)$.

(g) Narrower.

13.37

(a) IQ $= 183.994 + 0.04222$ (Weight) $- 3.909$ (Height) $+ 0.0002148$ (MRIcount).

(b) R Square $= 0.533$. About 53.3% of the variability in IQ can be explained by the explanatory variables Weight, Height, and MRIcount in the multiple linear regression model.

(c) The estimate of the population standard deviation is 18.46.

(d) Each t-test statistic is for testing whether the particular regression coefficient is equal to zero. The p-values for Height and MRIcount are < 0.05, so they are statistically significant at the 0.05 level. Weight however is not statistically significant at the 0.05 level.

13.39

(a) The scatterplot is provided below. Estimated correlations will vary here, should be positive.

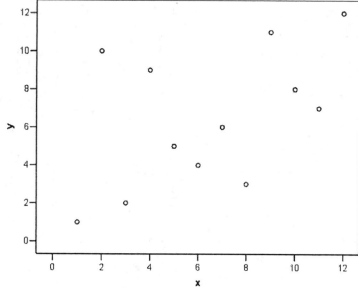

(b) $r = 0.497$.

(c) Answers will vary.

13.41

(a) There seems to be a strong positive linear relationship since $r = 0.993$ is close to 1. However, a scatterplot should be examined to assess the linear model assumption.

(b) $\hat{y} = 4.162 + 15.509(2) = 35.18$.

(c) Residual $= e = 29 - 35.18 = -6.18$.

(d) Unchanged.

(e) Outlier.

13.43

(a)

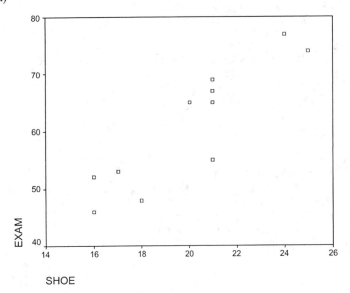

(b) $\hat{y} = -1.781 + 3.145 \, x$.

(c) For a one shoe size increase, we would expect an exam score increase of 3.145 points on average.

(d) $\hat{y} = -1.781 + 3.145(23) = 70.554$ points.

(e) Age (or grade level).

13.45

(a) Predicted earnings $\hat{y} = 375337.492 - 11940.924(10) = \$255,928.25$.

(b) $r = -0.765$.

(c) (i) Unchanged.
    (ii) Unchanged.

13.47

(a) $r = -0.857$.

(b) $\hat{y} = 20.56 - 0.399 \, x$.

(c) $\hat{y} = 20.56 - 0.399(42.2) = 3.72$.

(d) $120,000 = (1.2)(100,000)$ so we expect to have $(1.2)(3.72) = 4.46$ that is 4 or 5 cases of melanoma.

(e) No, 61.1 is outside the range of the x – values used to calculate the regression equation.

(f) (i) $r = 0.857$.
    (ii) slope $= 0.399$.

(g) Residual Plot

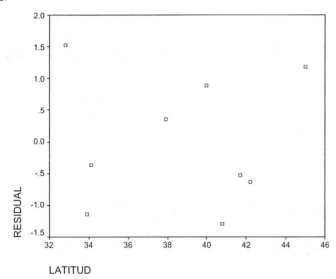

LATITUD

(h) Yes, the residuals form a random scatter around the zero-line.

13.49

(a) False.

(b) False.

13.51

(a) $b = r \left(\dfrac{s_Y}{s_X} \right) = 0.5251 \left(\dfrac{7.554}{3.392} \right) = 1.169$;

$a = \overline{y} - b\,\overline{x} = 1.878 - 1.169(1.304) = 0.3536$

so the equation is: $\hat{y} = 0.3536 + 1.169(x)$

And $r^2 = (0.5251)^2 = 0.2757 \rightarrow 27.57\%$

(b) Answers will vary.

13.53

(a) $\hat{y} = 321.241328 - 5.202654x$.

(b) The estimated slope of $b = -5.202654$ tells us that for one kg/day increase in total milk production for each cow, we would expect, on average, a decrease of 5.202654 uric-acid concentration in the milk.

(c) Yes. The p-value = 0 is given by Sig T.

13.55

(a) $\hat{y} = 70.717499 - 0.115611x$.

(b) 60 years old.

(c) $r = -0.80381$.

(d) Always.

(e) The p-value for assessing if there is a significant non-zero linear relationship is given in the output as 0.0000 in the row with variable PERTV. Since this is so small we would conclude there appears to be a significant linear relationship between life expectancy and number of people per TV.

13.57

(a) Negative.

(b) $\hat{y} = 23.354 - 8.168\,x$.

(c) Residual $= e = 15 - 18.126 = -3.126$.

(d) The 95% confidence interval for β is: $-8.168 \pm 2.228(2.061) \Rightarrow (-12.76,\ -3.576)$. Since 0 is not in the interval, we reject H_0.

(e) $\hat{y} = 23.354 - 8.16(0.750) = 17.234$.

(f) $\sqrt{7.043\left(\dfrac{1}{12} + \dfrac{(0.75 - 1.13)^2}{1.66}\right)} = 1.095$

(g) 61.1%.

14.1
(a) 7.779
(b) 9.488
(c) 13.277
(d) 1.064
(e) 10.645
(f) 18.549

14.3
(a) With the TI: p-value = 0.3711; Using Table VI: $0.10 < p$-value < 0.90.
(b) With the TI: p-value = 0.0845; Using Table VI: $0.05 < p$-value < 0.10.
(c) With the TI: p-value = 0.0005; Using Table VI: p-value < 0.005.

14.5
(a) 3.

(b) $\sqrt{2(3)} = 2.45$.

(c) $\frac{15-3}{2.45} = 4.9$

(d) Reject the null hypothesis since the observed test statistic value is nearly 5 standard deviations above what would be expected if the null hypothesis were true.
(e) The p-value is very small, about 0.0018. Thus we would reject the null hypothesis at the 5% level.

14.7
(a) $H_0 : p_A = p_B = p_C = p_D = p_E = 0.20$
(b) The five expected counts are all $(200)(0.20) = 40$.

(c) $X_{OBS}^2 = \frac{(38-40)^2}{40} + \frac{(44-40)^2}{40} + \frac{(56-40)^2}{40} + \frac{(37-40)^2}{40} + \frac{(25-40)^2}{40} = 12.75$.

(d) With 4 degrees of freedom, the p-value is $P(X^2 \geq 12.75) = 0.0126$ (using the TI), or using Table VI we have $0.01 < p$-value < 0.05.
(e) Since the p-value is less than 0.05, we reject H_0 and conclude that the five pain relief medicines do not appear to be equally effective.

14.9
(a) $H_0: p_1 = 0.25 \quad p_2 = 0.50 \quad p_3 = 0.25$
(b) $E_1 = E_3 = (0.25)(144) = 36$ and $E_2 = (0.50)(144) = 72$

$X^2 = 2.25 + 0.125 + 1 = 3.375$
df = 2; $0.15 < p$–value < 0.20 (or p–value = 0.185 using the TI)
So we fail to reject H_0 and conclude that the professor's theory appears to be supported by the data.

14.11

The observed test statistic is $X_{OBS}^2 = \frac{(16-19.2)^2}{19.2} + \frac{(23-21.6)^2}{21.6} + \frac{(9-7.2)^2}{7.2} = 1.074$. With 2 degrees of freedom, the p-value is $P(X^2 \geq 1.074) = 0.5845$ (with the TI) or $0.10 < p$-value < 0.90 (using Table VI). Since the p-value is greater than 0.01, so we fail to reject H_0 and conclude the distribution for domestic cars appears to be consistent with the distribution for non-domestic cars. Note: Since we observed a test statistic of 1.074 which is even less than what we'd expect to see if H_0 is true (df=2), it is certainly not large enough to reject H_0.

14.13

(a) About 18% of 1228 or about 221.

(b) About 28% of 1228 or about 344.

(c) Yes, at most 221.

(d) No, the categories are not disjoint (mutually exclusive), the percentages total to over 100%.

(e) The margin of error was approximately 0.029.

(f) Is it a random sample? Where was the survey conducted? Were the youth all selected from the same region or state? What kind of survey was it?

14.15

(a) $H_0: p_1 = 0.62, p_2 = 0.31, p_3 = 0.07$

(b) $800(0.07) = 56$ crimes.

(c) Theft: 496; Violent: 248.

(d) A chi-square distribution with 2 degrees of freedom.

(e) $X^2 = 8.96$.

(f) p-value $= 0.0113$. Using the table we have: $0.01 < p$-value < 0.025.

(g) We would reject the null hypothesis and conclude that there appears to have been a significant change in the distribution of school violence as compared to 10 years ago.

14.17

(a) The expected count for standard drug patients showing no change: $(100)(35)/210 = 16.67..$

(b) The observed test statistic is 1.76. With 2 degrees of freedom, the p-value is $P\left(X^2 \geq 1.76\right) = 0.416$ (with the TI) or $0.1 < p$-value < 0.9 (using Table VI). Thus our decision is to fail to reject H_0 and conclude that the distribution for patient condition appears to be the same for the two treatments.

14.19

(a) H_0: The distribution of salary for union members is the same as the distribution of salary for non-union members.

H_1: The distribution of salary for union members is not the same as the distribution of salary for non-union members.

(b) $Expected = \dfrac{20(30)}{70} = 8.57$

(c) $X^2 (2)$ distribution.

(d) Test statistic: 14.998; p-value $= 0.001$; Reject H_0 and conclude the distribution of salary does not appear to be the same for union members and non-union members.

14.21

(a) H_0: The three populations are homogeneous with respect to the distribution of alcohol usage, versus H_1: The three populations are not homogeneous with respect to the distribution of alcohol usage.

(b) Reject H_0 if the p-value is less than 0.05.

(c) You need to know the sample sizes.

(d) $X^2 = \dfrac{(34-50)^2}{50} + \dfrac{(48-50)^2}{50} + \dfrac{(68-50)^2}{50} + \dfrac{(66-50)^2}{50} + \dfrac{(52-50)^2}{50} + \dfrac{(32-50)^2}{50} = 23.36.$ The p-value is $P\left(X^2 \geq 23.36\right) = 0.0000085$ (with the TI) or p-value < 0.005 (using Table VI). Thus we reject H_0 and conclude that the distribution of alcohol usage does not appear to be the same for the 3 populations.

14.23

(a) H_0: Gender and Dessert Preference are independent for this population of students.
H_1: Gender and Dessert Preference are not independent for this population of students.

(b) The expected counts are: Ice Cream and Female = 26.8, Ice Cream and Male = 18.2
Yogurt and Female = 23.2, Yogurt and Male = 15.8.

(c) $X^2 = 6.65$

(d) The p-value is 0.0099, so we would reject the null hypothesis and conclude that there does appear to be a relationship between gender and dessert preference in this population of students.

14.25

(a) H_0: Big toe status and Roll tongue status are independent versus H_1: Big toe status and Roll tongue status are not independent.

(b) The expected number of people is $\dfrac{50(60)}{100} = 30$.

(c) The observed test statistic is 6.00, with a p-value of 0.01431. So our decision is to reject H_0. There does appear to be an association between Big Toe Status and Roll Tongue Status.

14.27

(a) H_0: Union membership and attitude are independent.

(b) The expected count for union members who are opposed is: $\dfrac{(176)(86)}{400} = 37.84$.

(c) The observed test statistic is 18.54498, with a p-value of 0.00009. Thus our decision is to reject H_0 and conclude that it appears that membership and attitude are associated.

(d) (i) H_0: Homogeneity, that is, the distribution of attitude is the same for the two populations.
(ii) We would reject H_0 since the p-value is less than 0.01.

14.29

(a) H_0: There is no relationship between age group and appearance satisfaction for the population of women.
H_1: There is a relationship between age group and appearance satisfaction for the population of women.

(b) $34/43 = 0.7907$.

(c) $(102)(43)/150 = 29.24$.

(d) The expected value is 2 (the degrees of freedom).

(e) The test statistic is 13.149 and the p-value is 0.001. Thus, there does appear to be an association between age group and appearance satisfaction.

14.31

(a) Proportion of men admitted: $90/200 = 0.45$ or 45%. Proportion of women admitted: $60/200 = 0.30$ or 30%. There appears to be an association between gender and admission status. The chi-square test of independence would give a test statistic of 9.6 and a p-value of 0.0019 (or using Table VI: p-value < 0.005). So there is strong evidence of an association between the two variables.

(b) Both departments admitted equally women and men applicants (50% in the engineering department, but only 25% in the English department). There does not appear to be any association between gender and admission status for either program. In fact the test statistic value for both programs would be 0 and the p-value = 1.

(c) Simpson's paradox or aggregation bias.

14.33

(a) The promotion rate for females of 260/1000 (or 26%) is higher than the promotion rate for males of 120/1000 (or 12%). A chi-square test homogeneity would yield an extremely large test statistic value of 63.68 and a p-value that is nearly 0. There is very strong support to say the incidence of promotion is not the same for male and female employees.

(b) Yes, as shown in the following possible example where females have a lower promotion rate at both job levels, but a higher proportion of females are in the job level the higher promotion rates overall.

Job Level 1	#	% promoted	# promoted
Males	100	30%	30
Females	900	28%	252

Job Level 2	#	% promoted	# promoted
Males	900	10%	90
Females	100	8%	8

14.35

(a) $H_0: p_1 = 0.25, p_2 = 0.50, p_3 = 0.25$, H_1: At least one of the equalities is not true.
where p_1 = proportion of the resulting offspring "AA" , p_2 = proportion of the resulting offspring "AB", p_3 = proportion of the resulting offspring "BB".

(b) We would expect about $0.25(188) = 47$.

(c) With 2 degrees of freedom, the p-value is $P(X^2 \geq 3.2) = 0.202$ (with the TI) or $0.10 < p$-value < 0.90 (using Table VI). So, we fail to reject H_0 for any $\alpha > 0.202$. These data do appear to support Mendel's theory.

14.37

(a) $H_0 : p_1 = \dfrac{1}{3}, p_2 = \dfrac{1}{3}, p_3 = \dfrac{1}{3}$ or $p_1 = p_2 = p_3 = \dfrac{1}{3}$

(b) $E_1 = E_2 = E_3 = (120)\left(\dfrac{1}{3}\right) = 40$

$$X^2 = \frac{(31-40)^2}{40} + \frac{(47-40)^2}{40} + \frac{(42-40)^2}{40} = \frac{81+49+4}{40} = 3.35$$

Using TI: the p-value $= 0.1873$. Using Table VI: $0.15 < p$-value < 0.20.
So we fail to reject H_0.

14.39

(a) H_0: The potency acceptance rates are the same for the 2 formulations, or the 2 formulations have homogeneous potency acceptance rates.

(b) The observed test statistic is 8, with a p-value of 0.00468 (with the TI) or p-value < 0.005 (using Table VI). Thus we reject H_0 and conclude that there appears to be a significant difference in the rates, with formulation 2 having the higher acceptance rate.

14.41

(a) The answer is (iii).

(b) Category 1 and Category 3: 200(0.25) = 50; Category 2: 200(0.50) = 100.

(c) The answer is (ii).

(d) The expected value is 2.

(e) No, the observed test statistic is exactly equal to the expected value for the test statistic under the null hypothesis. We also note that the p-value is between 0.25 and 0.50 which is larger than 0.10.

14.43

(a) The completed table is given by:

		District Type		
		Rural	Suburban	Urban
	Good Grades	38.3%	57.6%	68.6%
Goal	Popularity	33.6%	27.8%	17.1%
	Athletic Ability	28.2%	14.6%	14.3%

(b) The conditional distribution of Goal given District Type.

(c) Yes, the percentage with a goal of good grades increases from about 39% to nearly 69% as you go from rural to urban.

(d) X^2_{OBS} = 18.56 is the observed test statistic with 4 degrees of freedom, the p-value is $P(X^2 \geq 18.56)$= 0.00096 (with the TI) or p-value < 0.005 (using Table VI). So our decision is to reject the null hypothesis, which means there appears to be an association between district type and goal.

14.45

(a) The conditional distribution of smoking status given treatment group is:

		Treatment Group		
		Group 1	Group 2	Group 3
	Nonsmoker	19.8%	25.0%	19.2%
Smoking	Ex-Smoker	27.1%	24.0%	24.9%
Habit	Smoker	53.1%	51.0%	55.9%

(b) Profile plot:

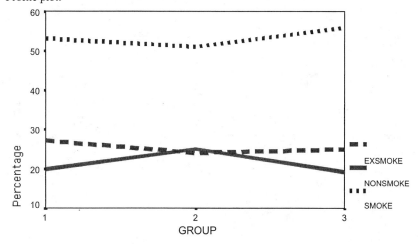

(c) The overall percentages are 21.4% nonsmokers, 25.3% ex-smokers, and 53.5% smokers.

(d) Enhanced profile plot:

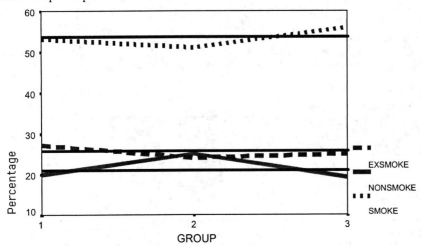

(e) Yes, the smoking habit rates are quite stable across the three groups.

14.47

(a) $H_0: p = 0.77$ versus $H_1: p > 0.77$.

(b) $z = \dfrac{\hat{p} - p_0}{\sqrt{\dfrac{p_0(1 - p_0)}{n}}} = \dfrac{\left(\dfrac{61}{72}\right) - 0.77}{\sqrt{\dfrac{0.77(0.23)}{72}}} = \dfrac{0.8472 - 0.77}{0.0496} = 1.56$

p-value = 0.0594.

(c) Yes, the p-value is less than 0.10.

(d) (i) $12/20 = 0.60$.
     (ii) $(20)(89)/120 = 14.83$.
     (iii) df = 3.
     (iv) The test statistic value is 15.1 and the p-value is 0.002.
     (v)   Yes, since the p-value is less than 0.05, we would reject H_0.

15.1

You might first look at the data with a histogram to assess if there appears to be strong skewness or outliers. A QQ plot of the data would also be a check on the assumption of normality required for the population or responses in order to perform the *t*-test.

15.3

(a) $R_{OBS} = 7$; *p*-value $= 0.1172 + 0.0439 + 0.0098 + 0.0010 = 0.1719 > 0.05$, fail to reject H_0.
(b) $R_{OBS} = 7$; *p*-value $= 0.9453 > 0.05$, fail to reject H_0.
(c) $R_{OBS} = 7$; *p*-value $= 2(0.1719) = 0.3438 > 0.05$, fail to reject H_0.
(d) The sample is selected randomly from a continuous population.

15.5

$H_0 : \pi_{0.5} = 258$ versus $H_1 : \pi_{0.5} > 258$

$R_{OBS} = 10$; *p*-value $= 0.0916 + 0.0417 + 0.0139 + 0.0032 + 0.0005 + 0.00003 = 0.15093 > 0.05$ so fail to reject H_0. The results are not statistically significant at the 5% level. There is not sufficient evidence to say the median round-trip ticket cost has increased.

15.7

$H_0 : \pi_{0.5}(populationA) = \pi_{0.5}(populationB)$ versus

$H_1 : \pi_{0.5}(populationA) \neq \pi_{0.5}(populationB)$.

$W = 66$; *p*-value $= 2(0.059) = 0.118$

Fail to reject H_0 and conclude that there is not a significant difference between the median scores for the two schools at the 10% level.

15.9

(a)

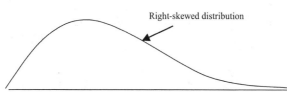

Right-skewed distribution

(b) $H_0 : \pi_{0.5}(population1) = \pi_{0.5}(population2)$ versus

$H_1 : \pi_{0.5}(population1) > \pi_{0.5}(population2)$ where 1 = Twin and 2 = Single.

(c) $W = 5 + 6 + 8 + 10 + 11 = 40$; *p*-value $= 0.041$
(d) Reject H_0.
(e) (i) Equal population standard deviations $\sigma_1 = \sigma_2$
    (ii) $t(9)$ distribution

15.11

(a) $H_0 : \pi_{0.5}(populationK) = \pi_{0.5}(populationG)$ versus

$H_1 : \pi_{0.5}(populationK) > \pi_{0.5}(populationG)$

(b) $W = 5 + 9 + 11 + 12 + 14 + 15 = 66$
(c) 48
(d) 8.49
(e) 2.12 standard deviations above
(f) 0.018
(g) The claim made by broker G, that its homes spend less time on the market, is supported by the data.

15.13

(a) $H_0: \mu_1 - \mu_2 = 0$ or $\mu_1 = \mu_2$ versus $H_a: \mu_1 - \mu_2 < 0$ or $\mu_1 < \mu_2$

(b) The observed test statistic is $t = -1.78$, with a p-value of $0.093/2 = 0.0465$. We would reject the null hypothesis and conclude that there is sufficient evidence (or it appears that) regular physical exercise leads to a lower resting pulse rate on average.

(c) (i) Wilcoxon Rank Sum Test.

(ii) $\mu_w = \dfrac{n_1(N+1)}{2} = \dfrac{(8)(20+1)}{2} = 84$

15.15

$H_0: \pi_{0.50}(D) = 0$ versus $H_1: \pi_{0.50}(D) > 0$ where $D = $ December $-$ January

$W^+ = 46$ and the p-value $= 0.032$ so we reject H_0. There is sufficient evidence to say the median number of baggage-related complaints has decreased overall.

15.17

(a) $H_0: \pi_{0.50}(D) = 0$ versus $H_1: \pi_{0.50}(D) > 0$ where $D = $ without $-$ with

(b) $W^+ = 11$ and the p-value $= 0.013$ so we reject H_0.

15.19

$H_0: \pi_{0.5} = 18$ versus $H_1: \pi_{0.5} \neq 18$

$R_{OBS} = 2$; p-value $= 2P(R \leq 2 \mid H_0$ is true$) = 2(0.1094 + 0.0313 + 0.0039) = 0.2892 > 0.05$ so we fail to reject H_0. The results are not statistically significant at the 5% level. The data do not refute the claim that the median percentage of chromium present is 18%.

15.21

(a) $t = 3/1.5 = 2$

(b) df $= 18$, p-value is 0.0608 or from Table IV: $0.05 < p$-value < 0.10.

(c) Wilcoxon rank sum test.

(d) $\mu_W = \dfrac{n_1(n_1 + n_2 + 1)}{2} = \dfrac{10(21)}{2} = 105$

15.23

(a) Signed rank test

(b) $W^+ = 3 + 4 + 5 + 6 = 18$ and the p-value $= 0.078$

(c) Reject H_0

(d) The Paris Weekend seems to be more popular than the Beach Package.

15.25

This is a paired design with two observations on each of the 10 subjects. The differences, computed as Placebo $-$ Actual Treatment are: 2, 0, 2, 6, 3, 1, -8, 5, 4, 2. The appropriate nonparametric test is the signed rank test.

The appropriate hypotheses are: $H_0: \mu_D = 0$ versus $H_1: \mu_D > 0$ where $D = $ placebo $-$ treatment.

There are $n^* = 9$ nonzero differences. The test statistic is $W^+ = 3 + 3 + 8 + 5 + 1 + 7 + 6 + 3 = 36$ and the p-value $= 0.064$. Using the 10% significance level, we would reject the null hypothesis and conclude that the treatment does appear to be effective at reducing the number of migraine attacks (as compared to placebo).

15.27

When we were able to use the pairing to control for patient-to-patient variability, the results were significant. Without being able to use this variance reducing feature, the same results would not be deemed statistically significant.